手機終端軟件開發實驗
ANDROID 版

羅文龍◎主編　蹇潔◎副主編

內 容 簡 介

隨著移動互聯網的快速發展,作為占據移動互聯網半壁江山的 Android 也迎來了更大的發展機遇,與其他行業類比,移動互聯網時代的 Android 發展潛力無疑是最大的(據 2016 年最新數據統計:Android 目前市場佔有率為 63.8%,iOS 市場佔有率為 19.1%)。

本書基於 Google 最新推出的 Android IDE - Android Studio 和最新的 Android 模擬器 Genymotion 作為開發環境進行編寫。本書全面介紹了 Android 應用程序開發相關知識的 8 大實驗環節,內容覆蓋了 Android 平臺搭建和 UI 設計、Andriod 高級 UI 設計、Intent 與 Activity 的使用、Android 資源訪問、圖形圖像與多媒體、Android 的網路編程基礎、SQLite 和 SQLiteDatabase 的使用、使用 GPS 與百度地圖等。

本書強調對 Android 相關知識的靈活應用,共包括 20 多個練習,每部分通過練習操作強化 Android 編程知識的學習。本書最後還提供了兩個綜合項目:基於 Android 的計算器、圖形化數字遊戲,綜合運用書中介紹的各項知識點,具有較高的參考價值。與本書配套的所有實例和綜合項目都可以登錄 https://yunpan.cn/c6H4UaNCUCCs6(提取碼:a2bb)進行免費下載。

本書可作為高等院校計算機科學與技術、軟件工程、信息管理、電子商務等相關專業本科生和研究生實驗環節教材,也可以供從事移動開發的工作者學習參考。

前 言

移動互聯網如潮水一般席捲著全世界，無論是個人還是企業、工作還是生活，都受其極大的影響。移動互聯網時代已經開啟，它已成為全世界商業和科技創新發展的加速器，成為當代最大的機遇和挑戰。

Android 系統就是一個開放式的移動互聯網操作系統。今天 Android 已經成為移動互聯網的寵兒，是應用最廣泛的移動互聯網平臺。因此手機軟件在當今的 IT 行業中具有舉足輕重的地位。從招聘市場的情況來看，Android 軟件人才的需求也越來越大。

在 2013 年 Google I/O 大會上，Google 正式推出了官方 Android 軟件集成開發工具 Android Studio，並在 2015 年宣布停止對 Android Eclipse Tools 的支持。以前很多書籍都是以 Eclipse 為開發環境進行編寫的，但以後 Android Studio IDE 開發必是大勢所趨，所以本書採用 Android Studio 作為練習開發平臺進行講解。

本書注重結合大學 Android 應用程序開發相關課程對應的實驗環節，突出與理論知識的結合性、實用性和可操作性，能夠使讀者在較短的時間內進行 Android 平臺搭建和 UI 設計、Andriod 高級 UI 設計、Intent 與 Activity 的使用、Android 資源訪問、圖形圖像與多媒體、Android 的網路編程基礎、SQLite 和 SQLiteDatabase 的使用、使用 GPS 與百度地圖等技術的熟練操作。

本教材具有以下特色：

1. 全新的開發環境

本書以最新的 Google 官方 Android IDE-Android Studio V1.3 為開發環境對練習進行開發講解，讓讀者更快地瞭解 Android Studio 的界面操作。同時還引入了當前應用廣泛的 Android 模擬器軟件 Genymotion，並對 Genymotion 的安裝、使用進行了詳細介紹，讓開發者擺脫 Android 模擬器運行緩慢、耗內存的缺點，使學習 Android 開發更加得心應手。

2. 由淺入深，緊扣理論課程

本書以高等教育本科學生為對象，緊扣 Android 理論知識教學環節，從瞭解 Android 和搭建開發環境學起，然後學習 Android 開發的基礎技術，進一步學習 Android 開發的高級內容，最後學習如何開發一個完整項目。講解過程中步驟詳盡、版式新穎，並在操作的內容圖片上進行了標註，讓讀者在閱讀時一目了然，從而快速掌握書中內容。

3. 知識全面、覆蓋面廣

書中全方面引入了 Android 的相關知識練習：Android 平臺搭建和 UI 設計、Andriod 高級 UI 設計、Intent 與 Activity 的使用、Android 資源訪問、圖形圖像與多媒體、

Android 的網路編程基礎、SQLite 和 SQLiteDatabase 的使用、使用 GPS 與百度地圖等。最後還提供了兩個綜合項目：基於 Android 的計算器、圖形化數字遊戲，綜合運用前面的各項知識點。

全書由羅文龍主編，蹇潔任副主編。羅文龍執筆編寫實驗一至實驗八，蹇潔執筆編寫綜合項目一、綜合項目二，全書由羅文龍負責審校和統稿。

本書僅基於 Android Studio V1.3 + API 18/23 + Genymotion 為開發環境進行講解，書中所論並不完美，錯誤和疏漏之處，懇請讀者批評指正。

編　者

目 錄

實驗一 Android 平臺搭建和 UI 設計 / 1
    1.1 Android 平臺搭建與 HelloWorld / 1
    1.2 簡單 UI 設計 / 31
    1.3 擴展練習 / 40
    1.4 實驗報告 / 41
    1.5 實驗成績考核 / 41

實驗二 Andriod 高級 UI 設計 / 42
    2.1 實驗目的 / 42
    2.2 實驗要求 / 42
    2.3 實驗內容 / 42
    2.4 擴展練習 / 54
    2.5 實驗報告 / 54
    2.6 實驗成績考核 / 55

實驗三 Intent 與 Activity 的使用 / 56
    3.1 實驗目的 / 56
    3.2 實驗要求 / 56
    3.3 實驗內容 / 56
    3.4 擴展練習 / 72
    3.5 實驗報告 / 72
    3.6 實驗成績考核 / 73

實驗四 Android 資源訪問 / 74
    4.1 實驗目的 / 74
    4.2 實驗要求 / 74
    4.3 實驗內容 / 74
    4.4 擴展練習 / 84

4.5 實驗報告　／84

4.6 實驗成績考核　／85

實驗五　圖形圖像與多媒體　／86

5.1 實驗目的　／86

5.2 實驗要求　／86

5.3 實驗內容　／86

5.4 擴展練習　／95

5.5 實驗報告　／95

5.6 實驗成績考核　／95

實驗六　Android 的網路編程基礎　／96

6.1 實驗目的　／96

6.2 實驗要求　／96

6.3 實驗內容　／96

6.4 擴展練習　／109

6.5 實驗報告　／109

6.6 實驗成績考核　／110

實驗七　SQLite 和 SQLiteDatabase 的使用　／111

7.1 實驗目的　／111

7.2 實驗要求　／111

7.3 實驗內容　／111

7.4 擴展練習　／132

7.5 實驗報告　／132

7.6 實驗成績考核　／132

實驗八　使用 GPS 與百度地圖　　/ 133
　　8.1　實驗目的　　/ 133
　　8.2　實驗要求　　/ 133
　　8.3　實驗內容　　/ 133
　　8.4　擴展練習　　/ 140
　　8.5　實驗報告　　/ 140
　　8.6　實驗成績考核　　/ 141

綜合項目一　基於 Android 的計算器　　/ 142
　　9.1　系統分析　　/ 142
　　9.2　系統設計　　/ 142
　　9.3　系統實施　　/ 144
　　9.4　系統運行與測試　　/ 157

綜合項目二　圖形化數字遊戲　　/ 159
　　10.1　系統分析　　/ 159
　　10.2　系統設計　　/ 159
　　10.3　系統實施　　/ 160
　　10.4　系統運行與測試　　/ 181

實驗一　Android 平臺搭建和 UI 設計

1.1　Android 平臺搭建與 HelloWorld

1.1.1　實驗目的

本次實驗的目的是讓大家熟悉搭建智能手機開發平臺的過程，瞭解 Android 開發項目的基本文件目錄結構，並實現 HelloWorld 小例程。

1.1.2　實驗要求

(1) 完成 Android 開發平臺的搭建及相關配置。
(2) 創建項目並熟悉文件目錄結構。
(3) 實現例程 HelloWorld。

1.1.3　實驗內容

1.1.3.1　安裝並配置 Java JDK

1. 下載 JDK

Java Development Kit(JDK)是 Sun 公司(2009 年 Sun 被 Oracle 收購)針對 Java 開發人員發布的免費軟件開發工具包(Software Development Kit,SDK)。自從 Java 推出以來,JDK 已經成為使用最廣泛的 Java SDK。作為 Java 語言的 SDK,普通用戶並不需要安裝 JDK 來運行 Java 程序,而只需要安裝 JRE (Java Runtime Environment)。而程序開發者必須安裝 JDK 來編譯、調試程序。下面以目前最新版本的 JDK 8 為例,介紹下載 JDK 的方法,具體步驟如下：

(1) 打開遊覽器,在地址欄中輸入 http://www.oracle.com/index.html,進入 Oracel 的官方主頁,如圖 1.1 所示。

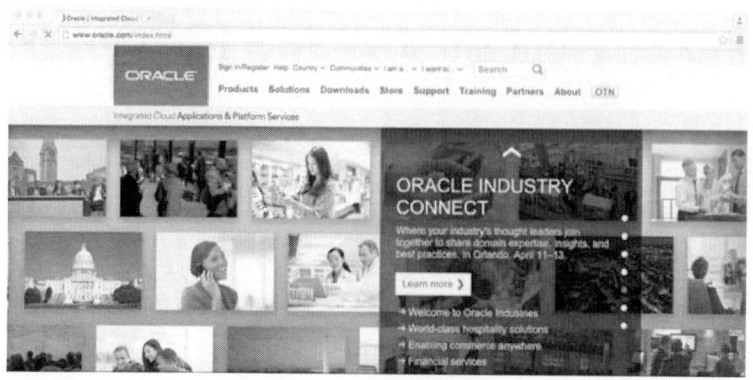

圖 1.1　Oracle 官方主頁

（2）選擇 Downloads 選項卡，選擇 Java for Developers，在跳轉的頁面，單擊 Java Platform（JDK）8u73／8u74 圖標，如圖 1.2 所示。

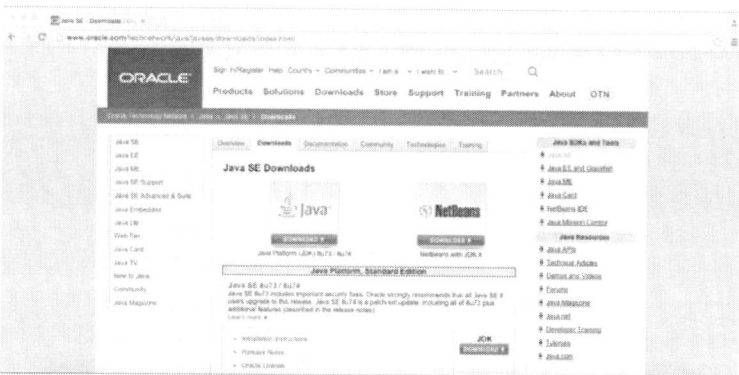

圖 1.2　Java for Developers 頁面

（3）在新頁面中，同意協議並根據計算機硬件和操作系統選擇適當的版本進行下載，如圖 1.3 所示。

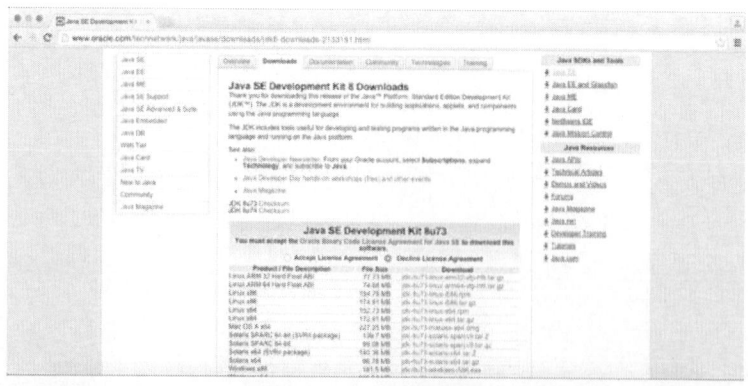

圖 1.3　JDK 下載頁面

2. JDK 的安裝

下載完適合自己操作系統的 JDK 版本以後,就可以進行安裝了。下面以 Windows 系統為例,講解 JDK 的安裝步驟:

(1)用鼠標左鍵雙擊 JDK 安裝包,會出現下圖所示的安裝界面,然後選擇「下一步」,如圖 1.4 所示。

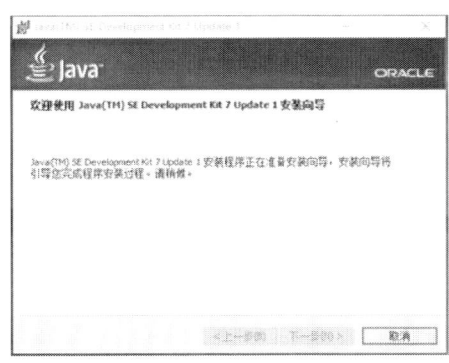

圖 1.4　JDK 安裝向導對話框

(2)在打開的如圖 1.5 所示的對話框中,單擊「更改」按鈕,將安裝位置改為 C：\Java\jdk1.7.0_01(根據版本決定最後 jdk 編號),如圖 1.6 所示。

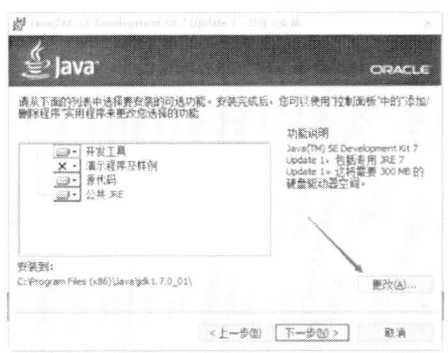

圖 1.5　JDK 安裝功能及位置選擇對話框(一)

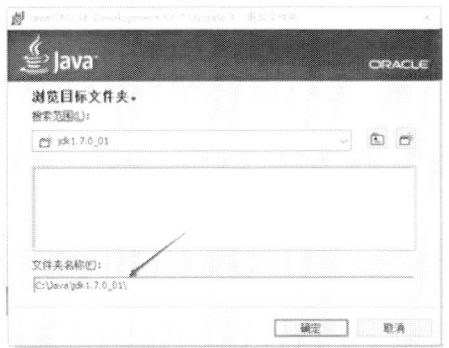

圖 1.6　JDK 安裝功能及位置選擇對話框(二)

(3)點擊「下一步」出現下圖所示的「正在安裝」界面,如圖 1.7 所示。

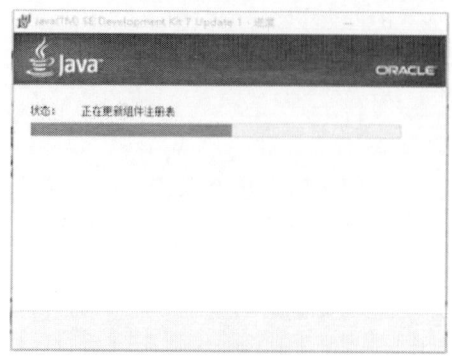

圖 1.7　JDK「正在安裝」界面

(4)當彈出如圖 1.8 所示的 JRE 安裝路徑選擇對話框時,單擊「更改」按鈕,將安裝路徑改為 C：\Java\jre7\,如圖 1.9 所示。

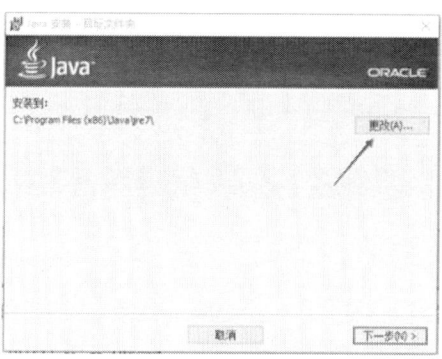

圖 1.8　JRE 安裝路徑選擇對話框(一)

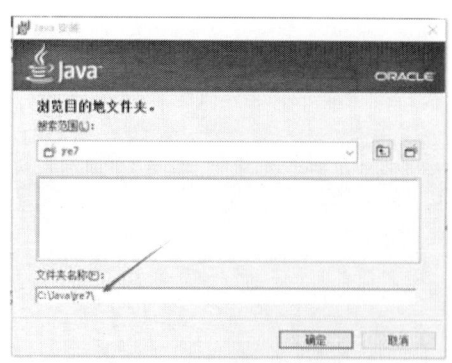

圖 1.9　JRE 安裝路徑選擇對話框(二)

(5)單擊「下一步」按鈕進行安裝,如圖 1.10 所示。

圖 1.10　JRE「正在安裝」介面

（6）安裝完成後，彈出如圖 1.11 所示的對話框，單擊「完成」，結束安裝。

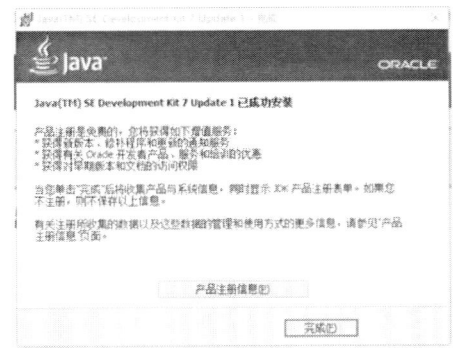

圖 1.11　JDK「完成安裝」界面

3. JDK 的環境變量配置

完成了前面的步驟，只是完成了 JDK 環境安裝。這個時候還要通過一系列的環境變量配置才能使用 JDK 環境進行 Android/Java 開發。配置環境變量包括 Java_home, path 和 classpath 三個部分。

（1）用鼠標右擊「我的電腦」，選擇「屬性」→「高級」→「環境變量」→「系統變量」→「新建」，如圖 1.12 所示。

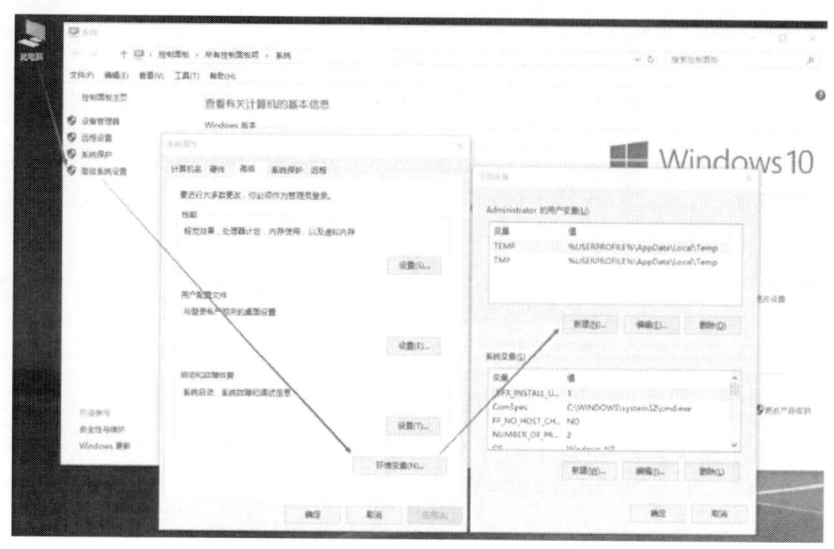

圖 1.12　Windows 環境變量對話框

（2）在「變量名」輸入框中寫入「Java_home」，在「變量值」輸入框中寫入「C：\Java\jdk1.7.0_01」（根據安裝路徑填寫），然後點擊「確定」，Java_home 就設置完成了。

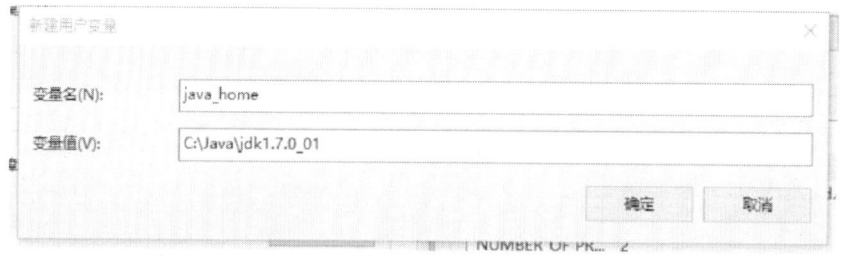

圖 1.13　配置 Java_home 變量

（3）下面開始「classpath」的配置。選中「系統變量」查看是否有 classpath 項目，如果沒有就點擊「新建」，如果已經存在就選中「classpath」選項，點擊「編輯」按鈕，然後在「變量名」中填寫「classpath」，在「變量值」中添寫「C：\Java\jdk1.7.0_01\jre\lib」（根據安裝路徑填寫）。注意：添加部分與前面用「；」隔開，如圖 1.14 所示。

圖 1.14　配置 Classpath 變量

（4）現在可以進行「Path」的配置了。同「classpath」設定類似，在「變量名」輸入框填寫「Path」，在「變量值」輸入框添加「C:\Java\jdk1.7.0_01\bin」（根據安裝路徑填寫）。注意：添加部分與前面用「；」隔開，如圖 1.15 所示。

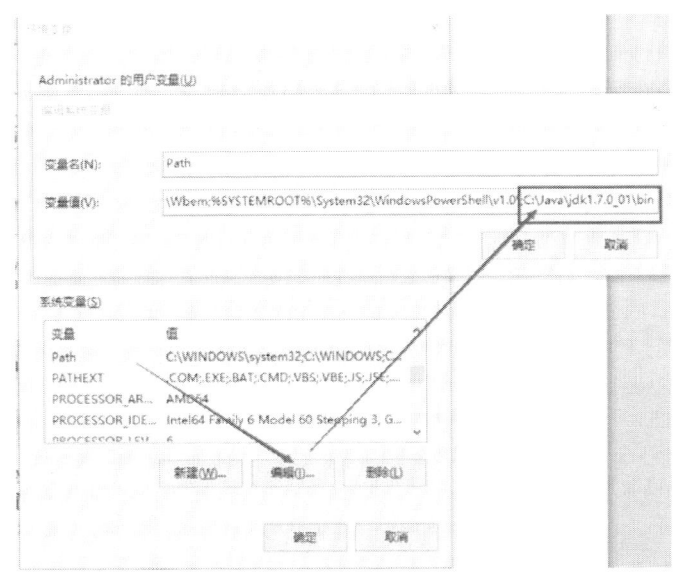

圖 1.15　配置 Path 變量

（5）JDK 的環境變量已經配置完成，可以通過打開命令提示符窗口，輸入命令「Java-version」，看到 Java 版本的信息，來確定安裝是否成功。首先點擊「開始」，然後點擊「所有應用程序」→「Windows 系統」→「命令提示符」，如圖 1.16 所示。

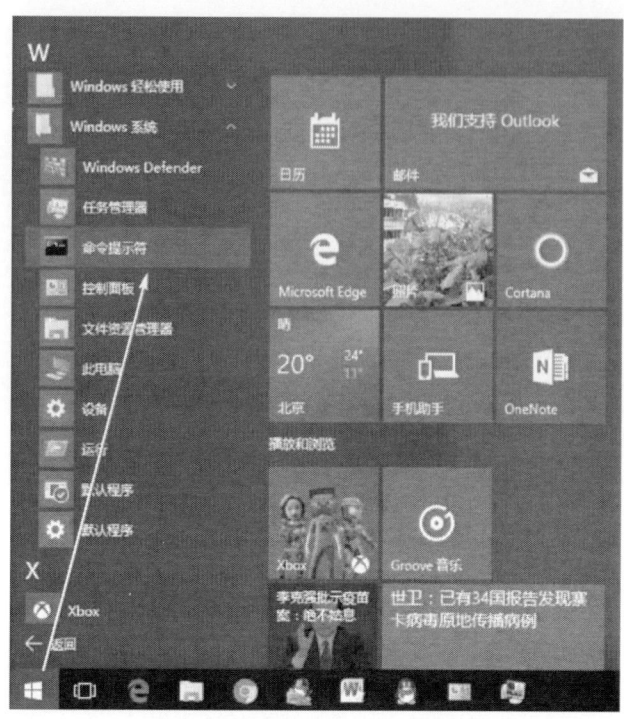

圖 1.16　Windows 中的命令提示符

（6）這個時候就進入了「命令提示符」窗口。在命令提示符窗口中輸入「Java-version」。注意 Java 和 -version 之間有一個空格，然後按「Enter」鍵，如圖 1.17 所示。

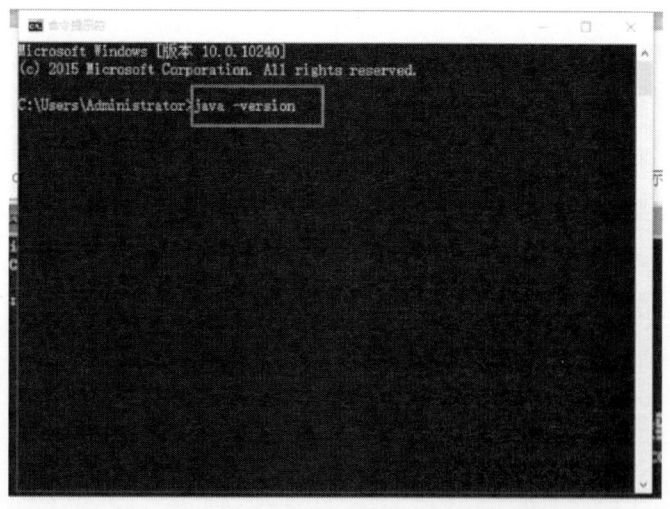

圖 1.17　「命令提示符」窗口運行「Java-version」命令

（7）JDK 版本信息就全部顯示出來了，也表明 JDK 已經安裝和配置完成，可以開始進行 Java 開發了，如圖 1.18 所示。

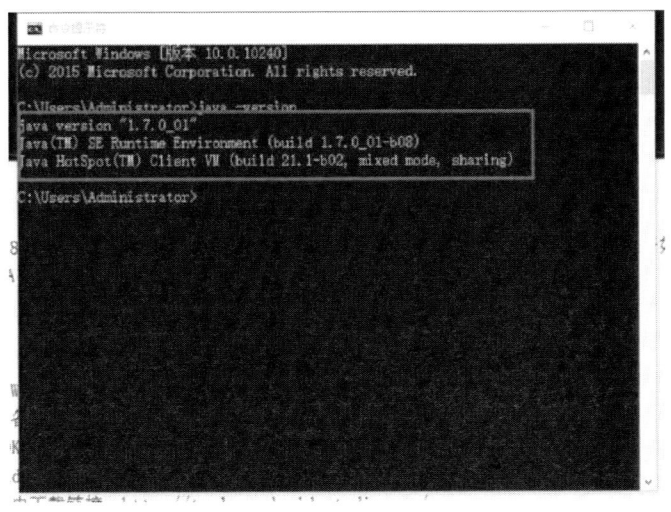

圖 1.18　JDK 版本信息

1.1.3.2　安裝 Android Studio

1. 準備工具

JDK 安裝包（JDK7 及以上版本）。

Android Studio 安裝文件。

國內下載連結：http://tools.android-studio.org/。

官網下載連結：http://developer.android.com/sdk/index.html。

2. 安裝文件

android-studio-bundle-141.2288178-windows.exe（1.1GB），推薦。

android-studio-ide-141.2288178-windows.exe（348MB），不包含 SDK Tools。

3. 說明

（1）32 位系統和 64 位系統使用同一個安裝文件。

（2）如果電腦中有 Android SDK，可以選擇不包含 SDK 的安裝版本。

（3）如果電腦已經安裝過 Android Studio，可以使用壓縮文件版本。

（4）建議使用包含 SDK 的安裝文件。

4. 安裝

這裡採用包含 SDK 的安裝文件進行講解，包含了不包含 SDK 的安裝文件的安裝步驟。如果你使用不包含 SDK 的安裝文件進行安裝，安裝步驟只會比這些步驟少而不會多。如果你使用壓縮包安裝，則可以直接跳過本節內容。

（1）找到下載的安裝文件，如圖 1.19 所示。

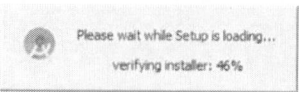

圖1.19　Android Studio 安裝文件

（2）雙擊安裝，如圖1.20 所示。

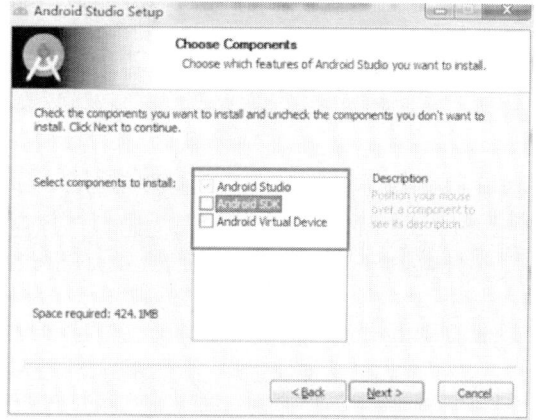

圖1.20　Android Studio 安裝解壓

（3）這裡我們不選擇下載SDK 和模擬器，如圖1.21 所示。

圖1.21　Android Studio 安裝向導

（4）選擇SDK 的安裝路徑，如圖1.22 所示。

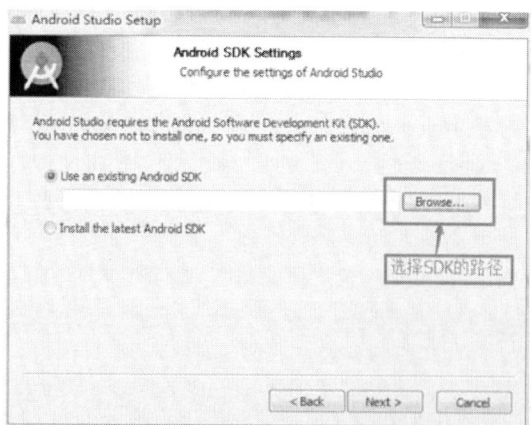

圖1.22　Android Studio 安裝路徑

(5)設置快捷方式,如圖 1.23 所示。

圖 1.23　Android Studio 安裝設置快捷方式

(6)安裝中,如圖 1.24 所示。

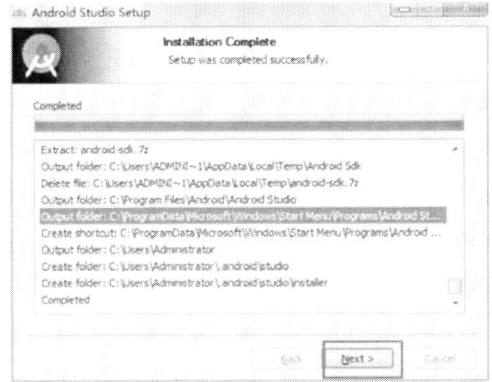

圖 1.24　Android Studio 安裝進行時

(7)安裝完成啓動 Android Studio,如圖 1.25 所示。

圖 1.25　Android Studio 安裝完成

(8)啓動 Android Studio,如圖 1.26 所示。

圖 1.26　Android Studio 啓動界面

(9)歡迎界面,如圖 1.27 所示。

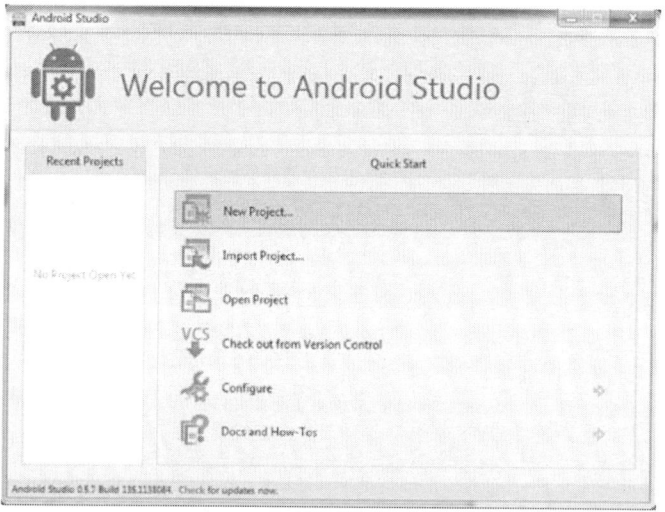

圖 1.27　Android Studio 歡迎界面

1.1.3.3　安裝 Android SDK

SDK Tools 其實就是 Android SDK Manager,即管理各種版本 SDK 的工具。Android SDK 包含模擬器、教程、API 文檔和示例代碼等內容。下面以 Windows 為例,詳細講解下載和安裝 Android SDK 的步驟:

(1)打開遊覽器,在地址欄中輸入 http://www.android-studio.org,進入 Android Studio 中文社區主頁,如圖 1.28 所示。

圖 1.28　Android Studio 中文社區主頁

(2)在主頁中找到 SDK TOOLS ONLY R24.3.4(當前最新版本),根據操作系統選擇安裝包進行下載,強烈建議下載主頁推薦版本「Installer_r24.3.4-windows.exe (Recommended)」,如圖 1.29 所示。

圖 1.29　SDK Tools 下載連接

(3)雙擊下載的程序,彈出如圖 1.30 所示的安裝向導對話框。

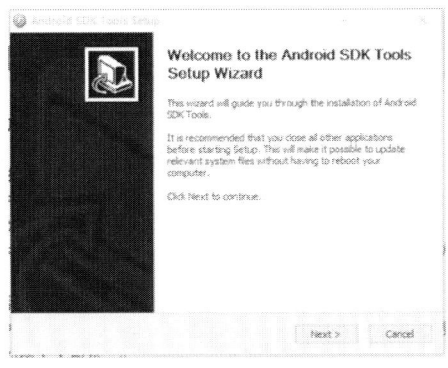

圖 1.30　SDK Tools 安裝向導

（4）單擊「Next」按鈕。如果已經正確安裝 JDK，則出現如圖 1.31 所示的界面。

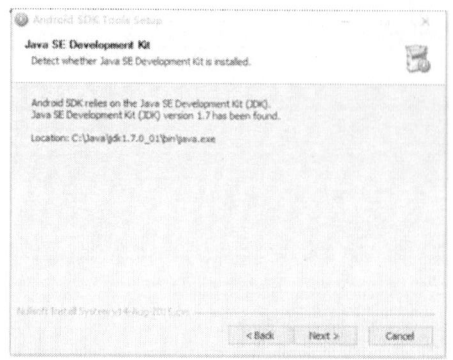

圖 1.31　SDK Tools 安裝配置 JDK 路徑

（5）在圖 1.31 中單擊「Next」按鈕，將提示選擇哪種用戶可以使用 SDK Tools，這裡選擇「Install for anyone using this computer」，如圖 1.32 所示。

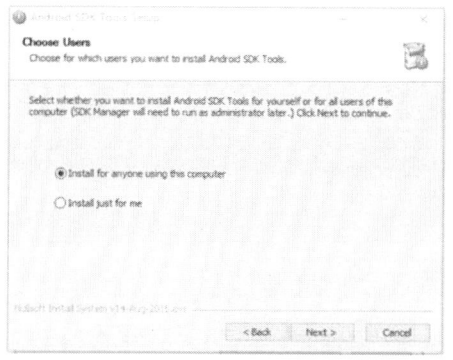

圖 1.32　SDK Tools 安裝配置用戶權限

（5）在圖 1.32 中單擊「Next」按鈕，將顯示 Android SDK 安裝路徑選擇窗口。將安裝路徑修改為「C：\Android\android-sdk」，如圖 1.33 所示。

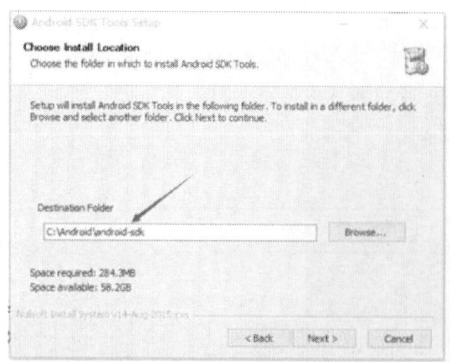

圖 1.33　SDK Tools 安裝配置安裝路徑

(6)在圖 1.33 中,單擊「Next」按鈕。此時詢問是否在開始菜單中創建快捷方式,如圖 1.34 所示。單擊「Install」按鈕開始安裝。

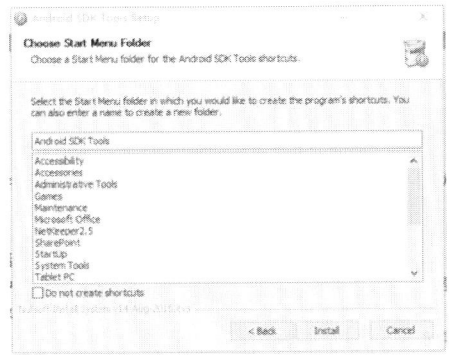

圖 1.34　SDK Tools 安裝配置菜單快捷方式

(7)這時開始安裝,如圖 1.35 所示。

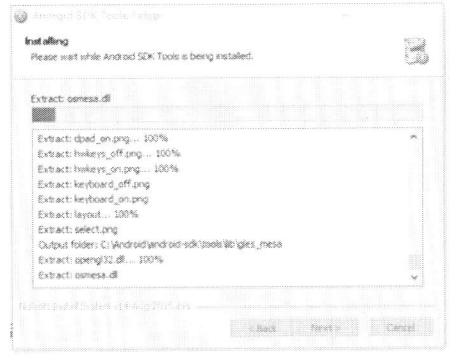

圖 1.35　SDK Tools 安裝進行時

(8)安裝完成後,顯示如圖 1.36 所示窗口,單擊「Next」按鈕。

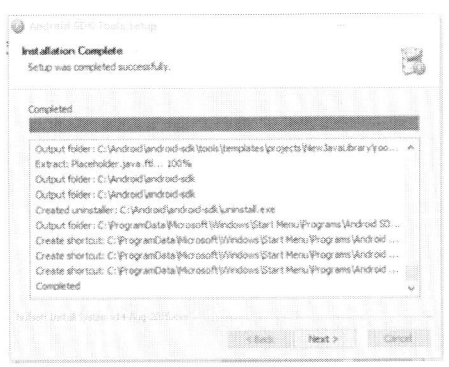

圖 1.36　SDK Tools 安裝完成(一)

(9)在圖1.37中,單擊「Finish」按鈕。

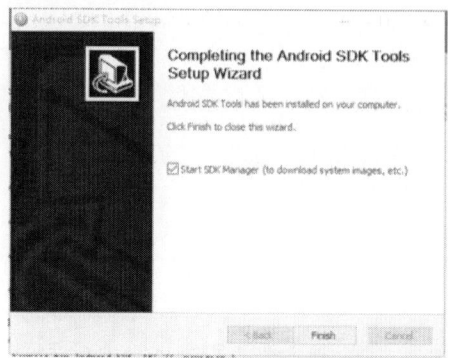

圖1.37　SDK Tools 安裝完成(二)

(10)啟動 SDK 管理工具。此時會自動聯網搜索可以下載的 API 等軟件包,如圖1.38所示。

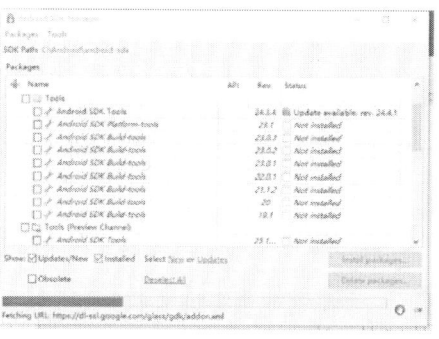

圖1.38　SDK Tools 運行界面

(11)為了便於今後在不同平臺中調試,在搜索完成後選擇安裝全部軟件包,如圖1.39所示。

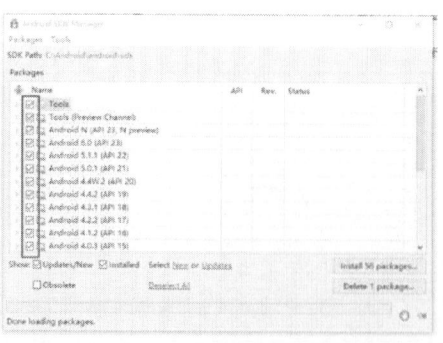

圖1.39　SDK Tools 選擇安裝 SDK 包文件

（12）在圖 1.39 中，單擊「Install packages」按鈕，安裝選中的軟件，如圖 1.40 所示。

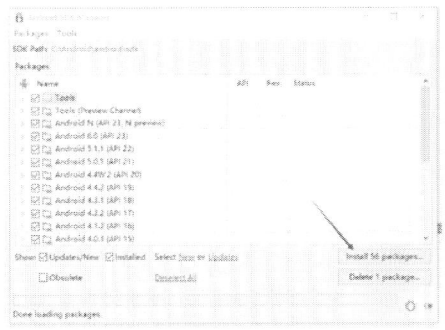

圖 1.40　在 SDK Tools 中安裝 SDK 包文件

（13）單擊「Install」按鈕，進行安裝。

1.1.3.4　安裝使用模擬器 Genymotion

1. 安裝模擬器 Genymotion

（1）雙擊運行下載的 Genymotion 安裝文件，選擇中文語言並點擊「Next」按鈕，如圖 1.41 所示。

圖 1.41　Genymotion 安裝向導

（2）可更改安裝路徑，點擊瀏覽「Browse..」按鈕，軟件默認的路徑為「C：\Program Files\Genymobile\Genymotion」，然後單擊「Next」按鈕，如圖 1.42 所示。

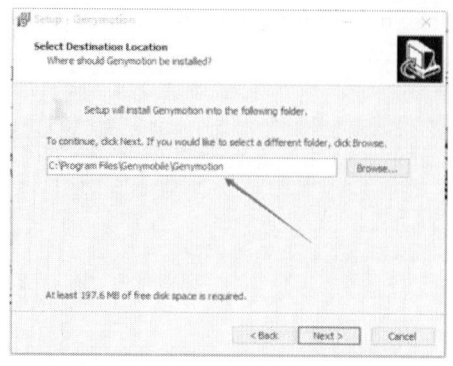

圖 1.42　Genymotion 安裝路徑對話框

（3）在彈出框選擇是否創建快捷菜單「Don't create a Start Menu folder」，然後點擊「Next」按鈕，如圖 1.43 所示。

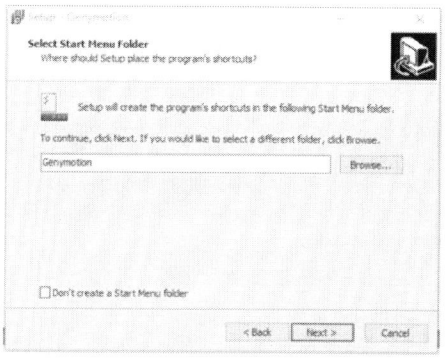

圖 1.43　Genymotion 安裝創建快捷菜單對話框

（4）在彈出框選擇是否創建桌面快捷方式「Create a desktop icon」，點擊「Next」→「Install」→「Finish」按鈕，如圖 1.44 所示。

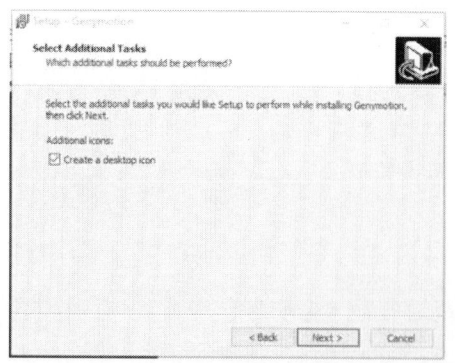

圖 1.44　Genymotion 安裝創建桌面快捷方式對話框

（5）在安裝完 Genymotion 後，會繼續安裝 VirtualBox。在 VirtualBox 安裝界面，

點擊「Next」按鈕,如圖 1.45 所示。

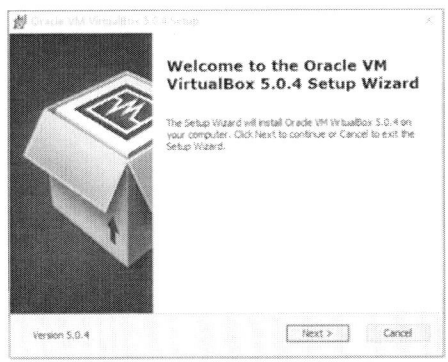

圖 1.45　VirtualBox 安裝向導對話框

(6)點擊瀏覽「Browse..」按鈕以更改 Location 的地址,VirtualBox 軟件默認路徑為「C:\Program Files\Oracle\VirtualBox\」,然後點擊「Next」按鈕,如圖 1.46 所示。

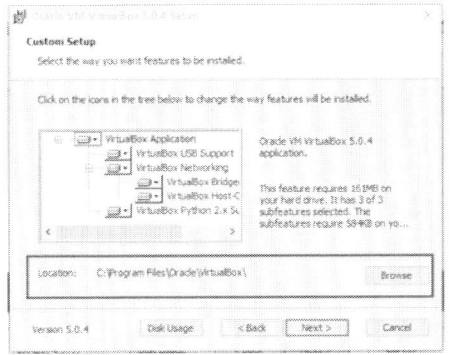

圖 1.46　VirtualBox 安裝更改安裝路徑對話框

(7)然後詢問是否現在安裝,選擇「Yes」按鈕,如圖 1.47 所示。

圖 1.47　VirtualBox 確認安裝對話框

(8)單擊「Install」開始安裝,然後單擊「Finish」按鈕完成安裝,如圖 1.48 所示。

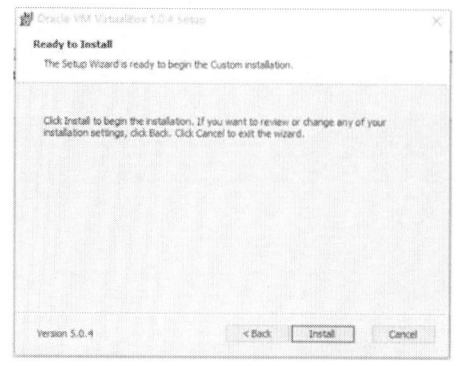

圖 1.48　開始安裝 Genymotion 對話框

2. 使用 Genymotion 模擬器

（1）第一次進入 Genymotion，系統會檢查你是否有安卓虛擬設備。如果沒有安裝則會彈出對話框，詢問你現在是否添加一個虛擬設備，如圖 1.49 所示，點擊「Yes」按鈕就可以了。

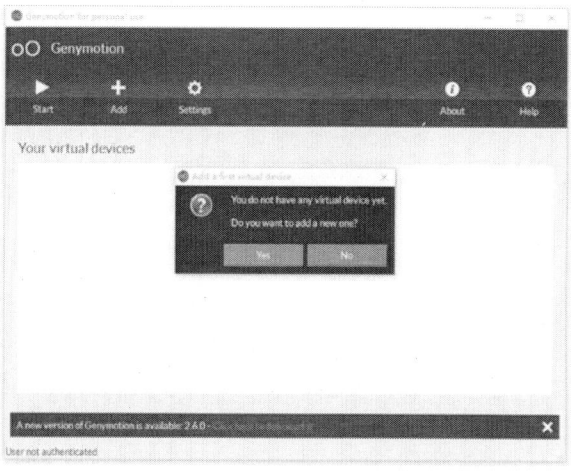

圖 1.49　第一次啓動 Genymotion 對話框

（2）創建一個新的虛擬設備「Create a new virtual device」。需要你輸入用戶名和密碼驗證。如：qq 郵箱××××××@qq.com（注意：如果驗證不通過，請到郵箱確認是否已經驗證過），如圖 1.50 所示。

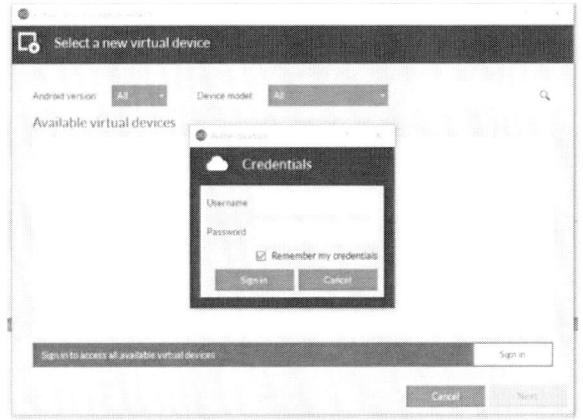

图 1.50　Genymotion 登录界面

(3)验证成功后,可以看到有很多虚拟设备,如:Samsung Galaxy S3,Samsung Galaxy S4 等。选择想添加的虚拟设备,选择后点击「Next」按钮,如图 1.51 所示。

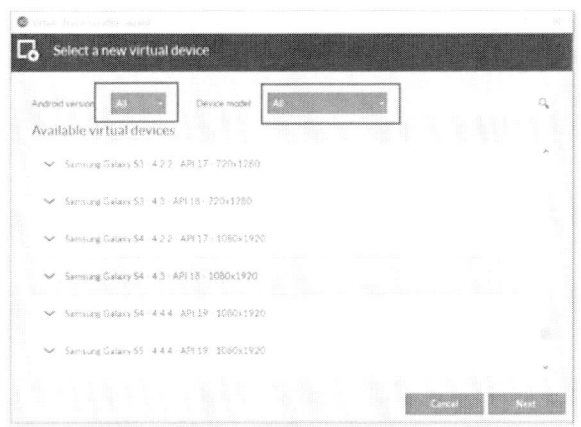

图 1.51　Genymotion 选择虚拟设备对话框

(4)下载安装,等下载到 100%,点击「Finish」按钮,如图 1.52 所示。

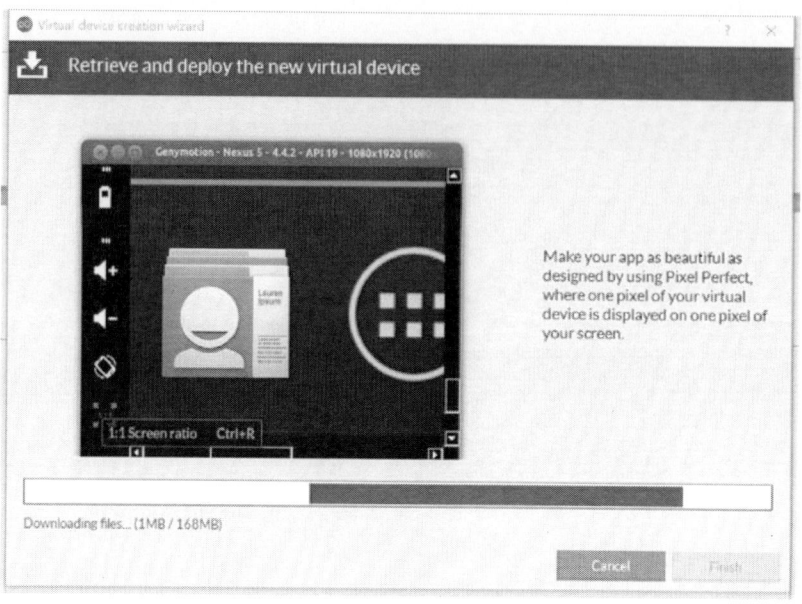

圖 1.52　下載模擬器進度對話框

（5）回到主窗口，選擇一個我們已經添加的模擬器，點擊啟動按鈕啟動模擬器，如圖 1.53 所示。

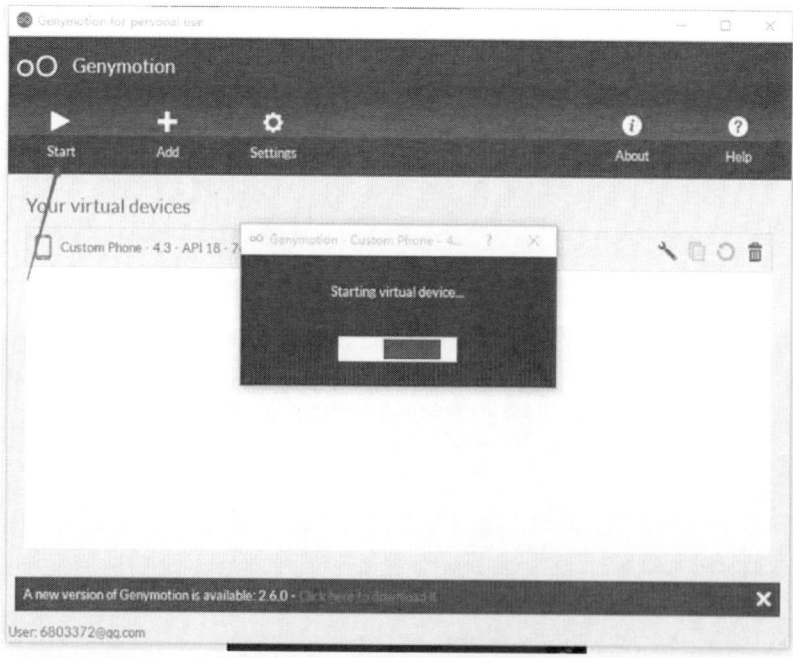

圖 1.53　Genymotion 加載對話框

(6)啟動虛擬機,將啟動如圖 1.54 的手機模擬界面。

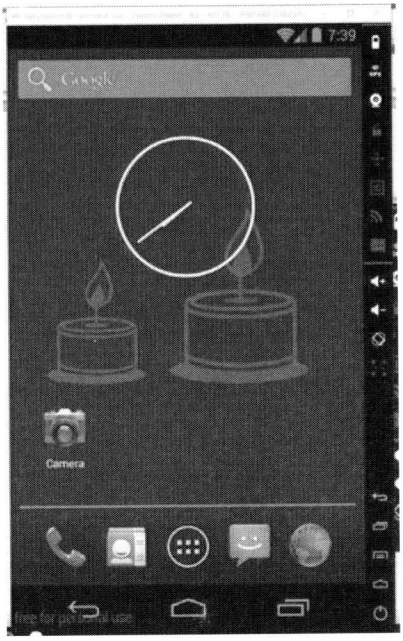

圖 1.54　Genymotion 虛擬機界面

3. Android Studio 安裝 Genymotion 插件

(1)打開 Android Studio,依次選擇菜單項「File」→「Settings」,如圖 1.55 所示。

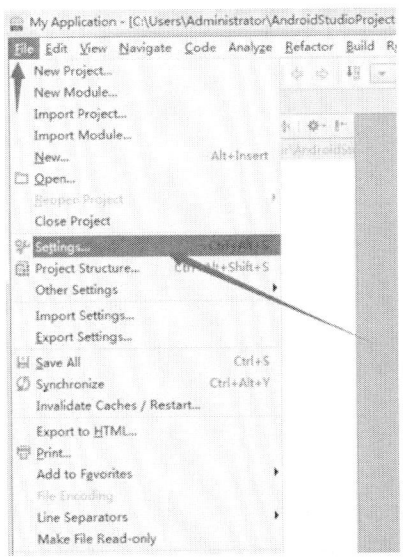

圖 1.55　Android Sutdio 文件菜單

（2）在打開的 Settings 對話框中找到「Plugins」設置項，單擊右側的「Browse repositories…」選項，如圖 1.56 所示。

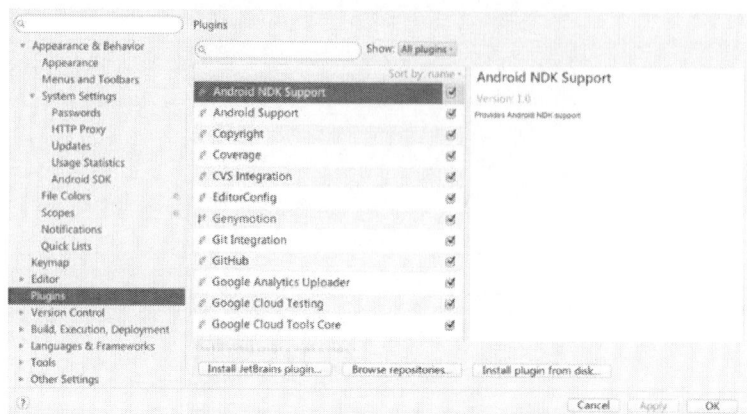

圖 1.56　Android Studio Plugins 設置對話框

（3）在搜索欄中輸入「genymotion」關鍵字，在右邊的框中將自動顯示已經搜索到的插件，單擊「Install plugin」安裝，如圖 1.57 所示。

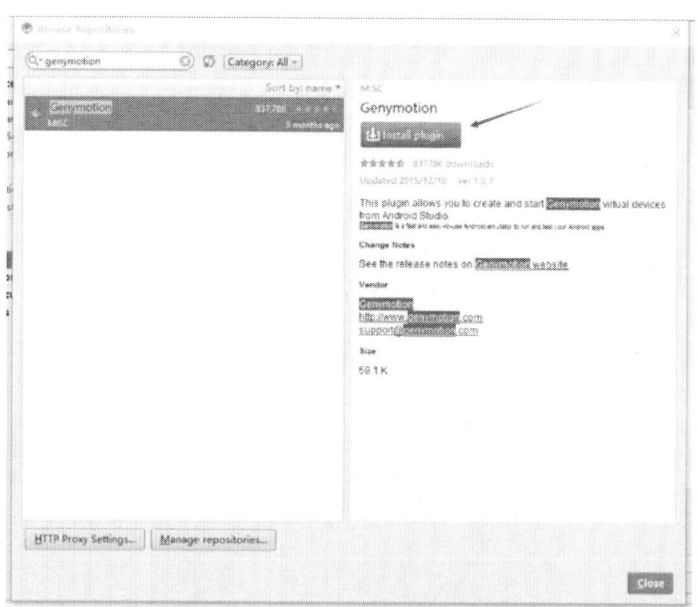

圖 1.57　安裝 Genymotion 插件對話框

（4）然後如圖 1.58 所示，開始下載。

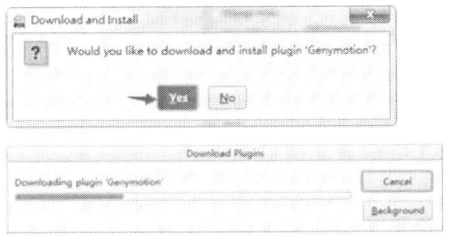

圖 1.58　Genymotion 下載界面

（5）安裝後重啟 Android Studio，在 Android Studio 的工具欄中將看到新增一個「▓」圖標，如圖 1.59 所示。

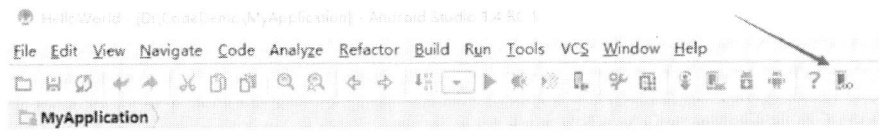

圖 1.59　Android Studio 工具欄

（6）初次單擊「▓」圖標還需要設置 Genymotion 的安裝目錄，如圖 1.60 所示。

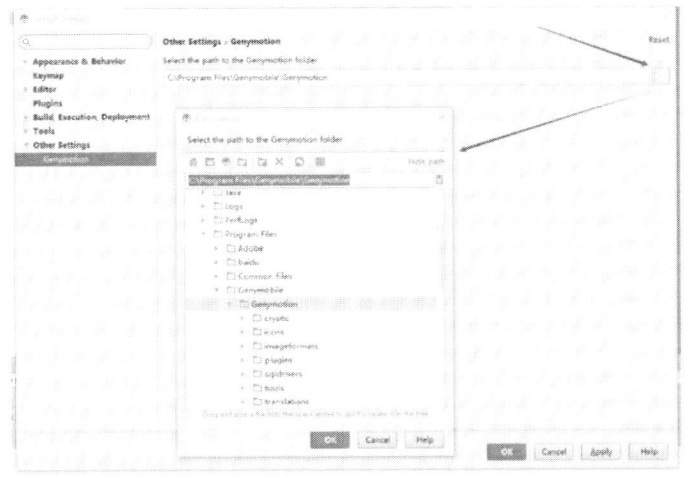

圖 1.60　設置 Genymotion 安裝目錄對話框

（7）設置好安裝目錄，再次單擊工具欄上的「▓」圖標就可以進行模擬器的設置和啟動，選中下載的模擬器然後單擊「Start…」以啟動模擬器，如圖 1.61 所示。

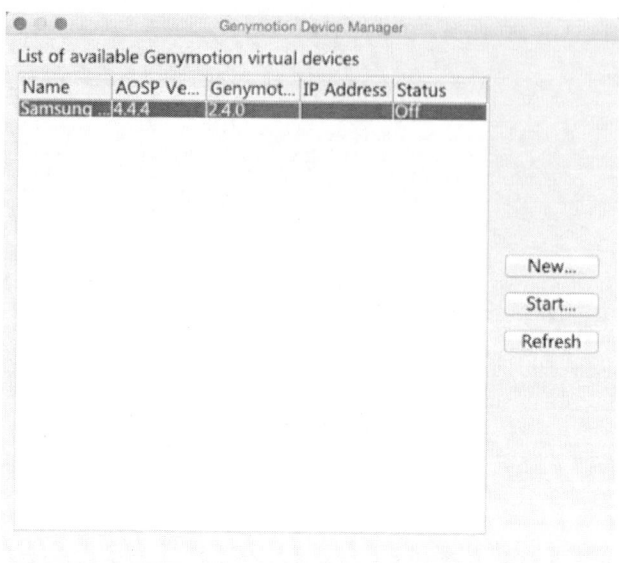

圖 1.61　Genymotion 啓動對話框

【練習 1.1】建立新項目 HelloWorld

配置好 JDK 和 Android SDK 後，就可以開始新建 Android 項目了。具體步驟如下：

（1）啓動 Android Studio，打開歡迎對話框，單擊「Start a new Android Studio project」選項，如圖 1.62 所示。

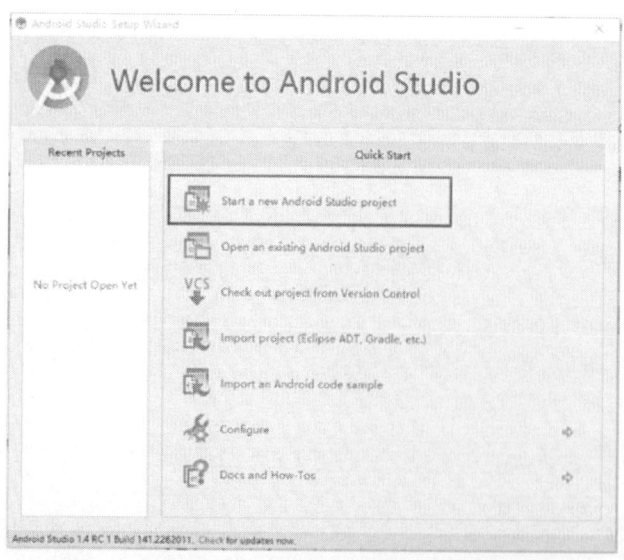

圖 1.62　在歡迎對話框中新建一個 Android Studio 項目

（2）在 Appication name 文本框中輸入「HelloWorld」，在 Company Domains 文本框中輸入「example.com」（根據自己需要改為姓名拼音.com 或公司英文名稱.com），在 Project location 文本框中選擇項目保存路徑，然後點擊「Next」按鈕，如圖 1.63 所示。

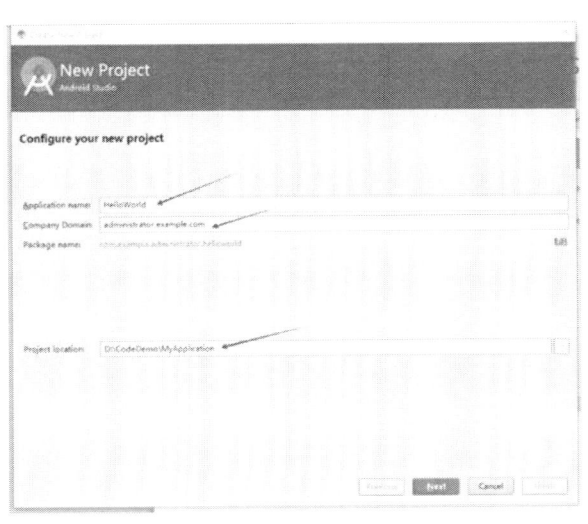

圖 1.63　設置新項目名稱對話框

（3）在彈出的對話框中選擇「Phone and Tablet」選項，「Minimum SDK」選項中選擇運行 Android 最低 SDK 版本要求[此處以 API:18 Android 4.3(Jelly Bean)為例]，然後單擊「Next」按鈕，如圖 1.64 所示。

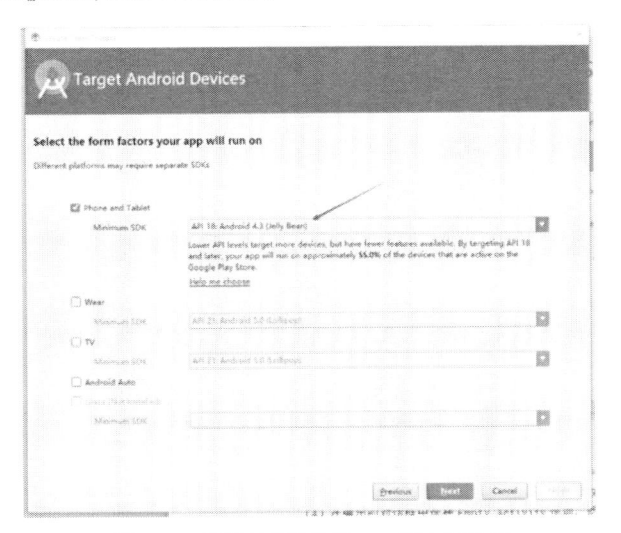

圖 1.64　設置新項目最低 SDK 版本對話框

（4）在彈出的對話框中選擇「Empty Activity」選項，然後單擊「Next」按鈕，如圖 1.65 所示。

图 1.65 选择 Activity 类型对话框

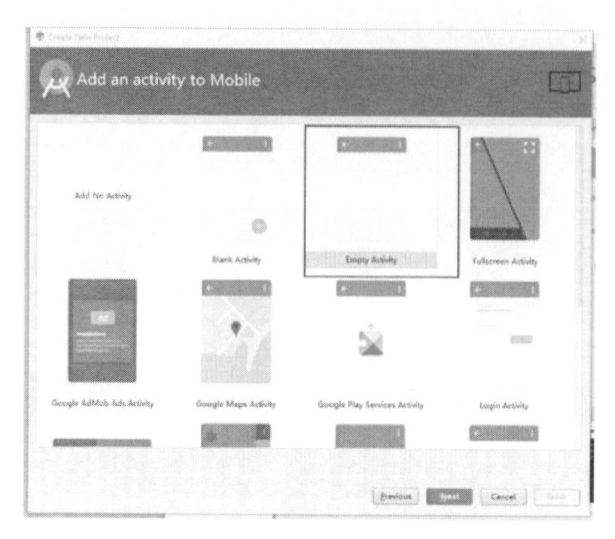

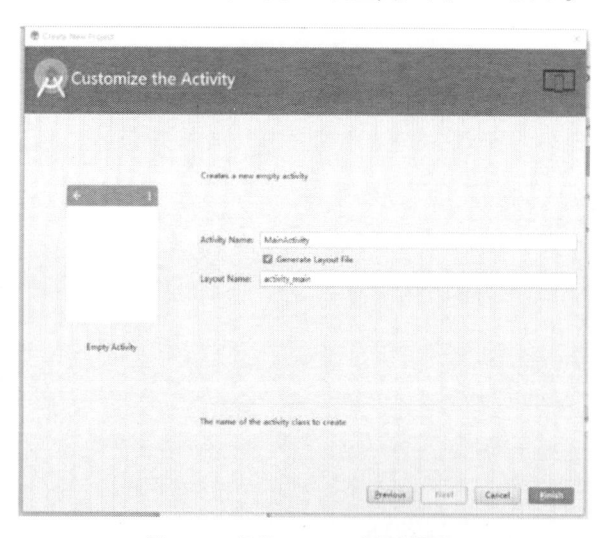

(5)在弹出的对话框中单击「Finish」按钮(此对话框将设置 Activity,Layout 的名称,建议初学者不要修改此对话框的默认设置),如图 1.66 所示。

图 1.66 设置 Activity 名词对话框

(6)加载新创建的 Android Studio 项目,会出现如图 1.67 所示窗口。

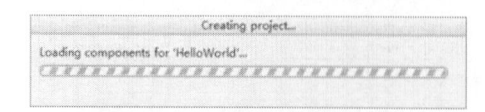

图 1.67 创建项目加载对话框

(7)创建好项目后会出现如图 1.68 的编辑界面。

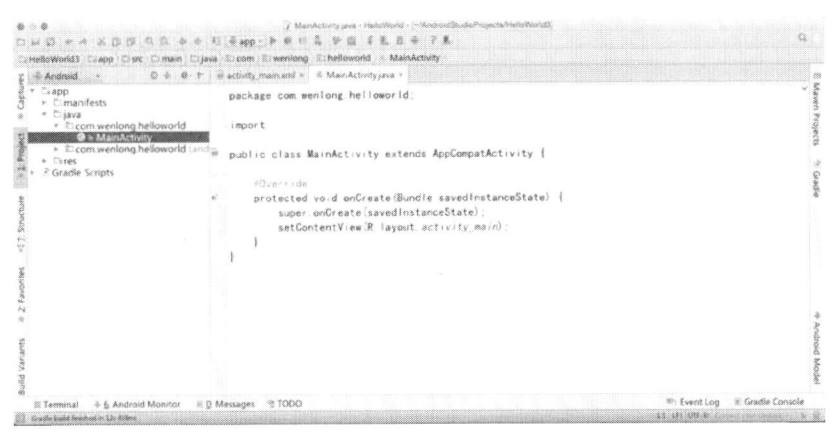

圖 1.68　Android Studio 編輯界面

(8) 啓動 Genymotion 模擬器：單擊 Android Studio 工具條中的「　　」按鈕，彈出如圖 1.69 的對話框。選擇一個已經下載的模擬器平臺，然後單擊「Start…」按鈕。

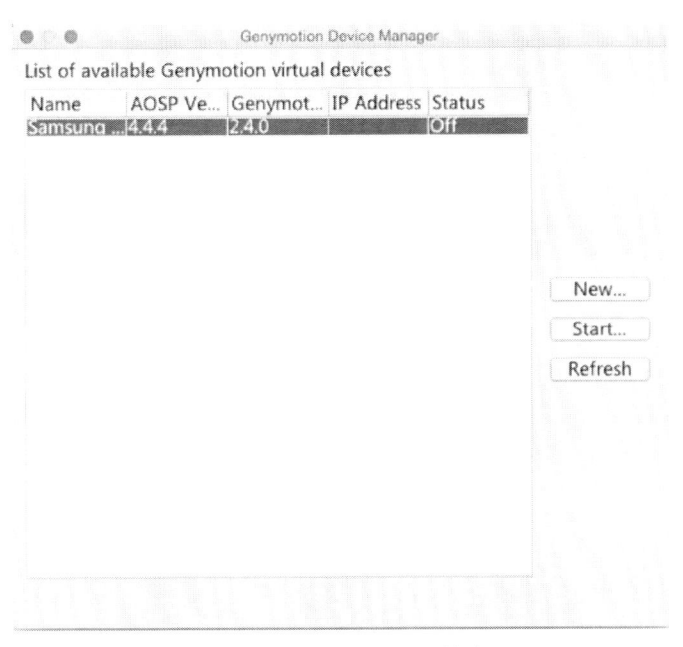

圖 1.69　選擇模擬器平臺對話框

(9) 模擬器平臺啓動成功後將彈出如圖 1.70 的手機模擬界面，滑動屏幕鎖以解鎖屏幕，如圖 1.71 所示。

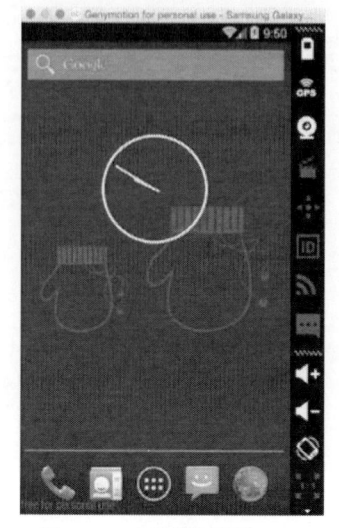

圖 1.70　手機模擬界面(一)　　　　　　圖 1.71　手機模擬界面(二)

（10）單擊 Android Studio 工具條中的「▶」按鈕，彈出如圖 1.72 所示的項目運行方式選擇對話框。確認有剛才運行的 Genymotion 模擬器，然後單擊「OK」按鈕。

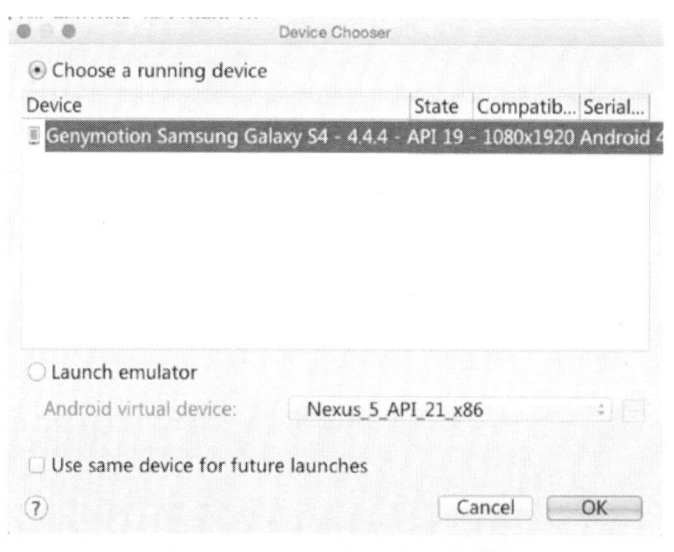

圖 1.72　Android Studio 項目運行方式選擇對話框

（11）在模擬器中查看 HelloWorld 項目的運行效果，如圖 1.73 所示。

圖 1.73　HelloWorld 項目運行效果圖

1.2　簡單 UI 設計

1.2.1　實驗目的

本次實驗的目的是讓大家熟悉 Android 開發中的 UI 設計,包括瞭解和熟悉常用控件的使用、界面佈局和事件處理等內容。

1.2.2　實驗要求

(1)熟悉和掌握界面控件設計。
(2)瞭解 Android 界面佈局。
(3)掌握控件的事件處理。

1.2.3　實驗內容

Android 中有許多常用控件,主要有文本框、按鈕和列表等。
文本框:TextView,EditText。
按鈕:Button,RadioButton,RadioGroup,CheckBox,ImageButton。
列表:List,ExpandableListView,Spinner,AutoCompleteTextView,GridView,Image-

View。

　　進度條：ProgressBar，ProgressDialog，SeekBar，RatingBar。

　　選擇器：DatePicker，TimePicker。

　　菜單：Menu，ContentMenu。

　　對話框：Dialog，ProgressDialog。

【練習1.2】運用ImageView

實現通過ImageView顯示帶邊框的圖片。

(1)運行效果如圖1.74所示。

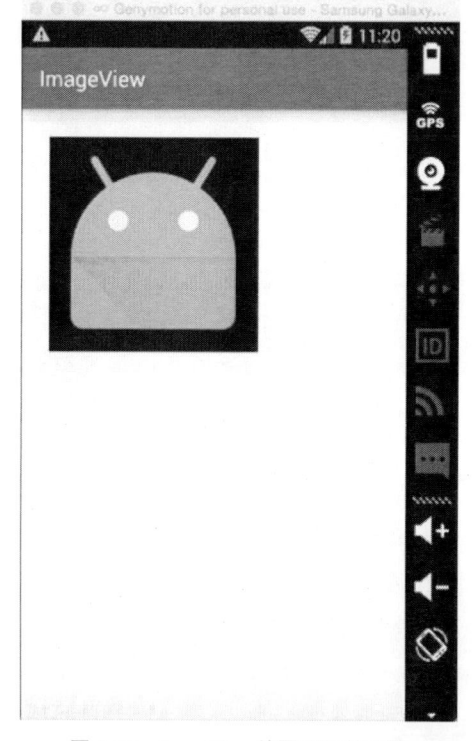

圖1.74　ImageView練習運行效果圖

(2)資源文件佈局如圖1.75所示。

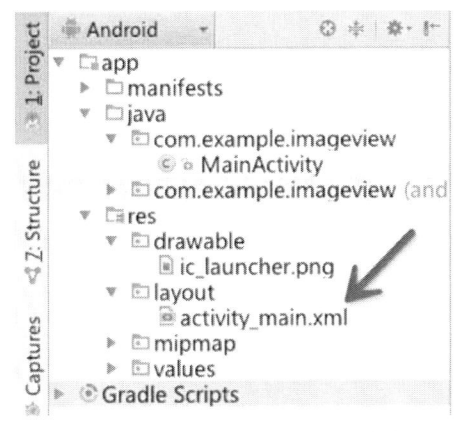

圖 1.75 ImageView 練習工程佈局結構圖

· activity_main.xml 文件代碼如下：

```xml
<? xml version="1.0" encoding="utf-8"? >
<RelativeLayout xmlns:android="http://schemas.android.com/apk/res/android"
    xmlns:tools="http://schemas.android.com/tools" android:layout_width="match_parent"
    android:layout_height="match_parent" android:paddingLeft="@dimen/activity_horizontal_margin"
    android:paddingRight="@dimen/activity_horizontal_margin"
    android:paddingTop="@dimen/activity_vertical_margin"
    android:paddingBottom="@dimen/activity_vertical_margin" tools:context=".MainActivity">

    <ImageView
        android:id="@+id/imageView1"
        android:padding="2dp"
        android:layout_margin="10dp"
        android:layout_width="wrap_content"
        android:layout_height="wrap_content"
        android:background="#000"
        android:src="@drawable/ic_launcher" />

</RelativeLayout>
```

其中，android:src="@drawable/ic_launcher" 為設置圖片資源語句。

【練習 1.3】運用 CheckBox

實現選中復選框後,「開始」按鈕才可用,否則為不可用狀態。
(1)運行效果圖如圖 1.76 所示。

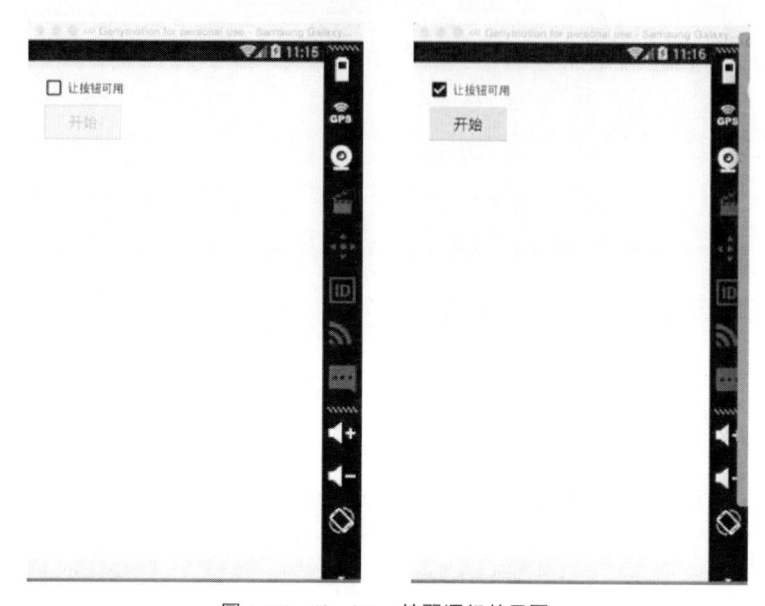

圖 1.76　CheckBox 練習運行效果圖

(2)資源文件佈局為如圖 1.77 所示。

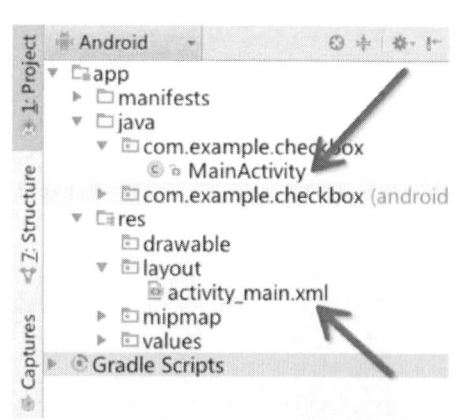

圖 1.77　CheckBox 練習工程結構圖

・activity_main.xml 文件代碼如下:

```xml
<?xml version="1.0" encoding="utf-8"?>
<RelativeLayout xmlns:android="http://schemas.android.com/apk/res/android"
    xmlns:tools="http://schemas.android.com/tools" android:layout_width="match_parent"
    android:layout_height="match_parent" android:paddingLeft="@dimen/activity_horizontal_margin"
    android:paddingRight="@dimen/activity_horizontal_margin"
    android:paddingTop="@dimen/activity_vertical_margin"
    android:paddingBottom="@dimen/activity_vertical_margin" tools:context=".MainActivity">

    <CheckBox
        android:id="@+id/checkBox1"
        android:layout_width="wrap_content"
        android:layout_height="wrap_content"
        android:text="让按钮可用" />
    <Button
        android:id="@+id/button1"
        android:layout_width="wrap_content"
        android:layout_height="wrap_content"
        android:enabled="false"
        android:text="开始"
        android:layout_below="@+id/checkBox1"
        android:layout_alignParentStart="true"
        android:layout_alignEnd="@+id/checkBox1" />

</RelativeLayout>
```

· MainActivity.java 文件代码为：

```java
package com.example.checkbox;

import android.app.Activity;
import android.os.Bundle;
import android.view.View;
import android.widget.Button;
import android.widget.CheckBox;
import android.widget.CompoundButton;
import android.widget.Toast;

public class MainActivity extends Activity {

    @Override
    protected void onCreate(Bundle savedInstanceState) {
        super.onCreate(savedInstanceState);
        setContentView(R.layout.activity_main);
        final Button button = (Button) findViewById(R.id.button1);
        button.setOnClickListener(new View.OnClickListener() {
            @Override
            public void onClick(View v) {
                Toast.makeText(MainActivity.this, "單擊了按鈕!", Toast.LENGTH_SHORT).show();
            }
        });
        CheckBox checkbox = (CheckBox) findViewById(R.id.checkBox1);
        checkbox.setOnCheckedChangeListener(new CompoundButton.OnCheckedChangeListener() {
            @Override
            public void onCheckedChanged(CompoundButton buttonView, boolean isChecked) {
                if (isChecked) {            //判斷該復選按鈕是否被選中
                    button.setEnabled(true);      //設置按鈕可用
                } else {
                    button.setEnabled(false);     //設置按鈕不可用
                }
            }
        });
    }
}
```

【練習 1.4】運用 ListView

編寫 Android 程序,實現圖標在上、文字在下的 ListView。
(1)運行效果如圖 1.78 所示。

圖 1.78　ListView 練習運行效果圖

(2)資源文件佈局如圖 1.79 所示。

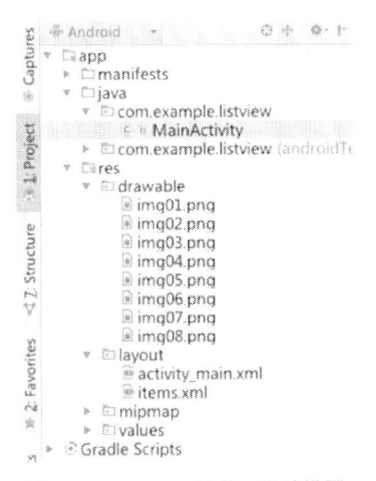

圖 1.79　ListView 練習工程結構圖

(3)圖標文件(img01.png-img08.png)如圖 1.80 所示。

圖 1.80　ListView 練習用到的圖標

- activity_main.xml 文件代碼如下：

```xml
<?xml version="1.0" encoding="utf-8"?>
<LinearLayout xmlns:android="http://schemas.android.com/apk/res/android"
    android:orientation="vertical"
    android:layout_width="fill_parent"
    android:layout_height="fill_parent"
    >
<ListView
        android:id="@+id/listView1"
        android:layout_height="wrap_content"
        android:layout_width="match_parent" />

</LinearLayout>
```

- items.xml 文件代碼如下：

```xml
<?xml version="1.0" encoding="utf-8"?>
    <LinearLayout
        xmlns:android="http://schemas.android.com/apk/res/android"
        android:orientation="vertical"

        android:layout_width="match_parent"
        android:layout_height="match_parent" >
```

```
<ImageView
        android:id = "@ +id/image"
        android:adjustViewBounds = "true"
        android:maxWidth = "72dp"
        android:maxHeight = "72dp"
        android:layout_height = "wrap_content"
        android:layout_width = "wrap_content" />
<TextView
        android:layout_width = "wrap_content"
        android:layout_height = "wrap_content"
        android:padding = "10dp"
        android:id = "@ +id/title"
        />
</LinearLayout>
```

· MainActivity.java 文件代碼為：

```
package com.example.listview;
    import android.app.Activity;

    import android.os.Bundle;
    import android.widget.ListView;
    import android.widget.SimpleAdapter;

    import java.util.ArrayList;
    import java.util.HashMap;
    import java.util.List;
    import java.util.Map;

    public class MainActivity extends Activity {
        /** Called when the activity is first created. */
        @Override
        public void onCreate(Bundle savedInstanceState) {
            super.onCreate(savedInstanceState);
            setContentView(R.layout.activity_main);
            ListView listview = (ListView) findViewById(R.id.listView1); // 獲取列表視圖
            int[] imageId = new int[] { R.drawable.img01, R.drawable.img02,
```

R.drawable.img03，R.drawable.img04，R.drawable.img05，R.drawable.img06，R.drawable.img07，R.drawable.img08 };// 定義並初始化保存圖片 id 的數組
　　　　String[] title = new String[]{"保密設置","安全","系統設置","上網","我的文檔",
　　　　"GPS 導航","我的音樂","E-mail"};// 定義並初始化保存列表項文字的數組
　　　　List<Map<String, Object>> listItems = new ArrayList<Map<String, Object>>();// 創建一個 list 集合
　　　　// 通過 for 循環將圖片 id 和列表項文字放到 Map 中,並添加到 list 集合中
　　　　for (int i = 0; i < imageId.length; i++){
　　　　　　Map<String, Object> map = new HashMap<String, Object>();// 實例化 Map 對象
　　　　　　map.put("image", imageId[i]);
　　　　　　map.put("title", title[i]);
　　　　　　listItems.add(map);// 將 map 對象添加到 List 集合中
　　　　}

　　　　SimpleAdapter adapter = new SimpleAdapter(this, listItems,
　　　　　　R.layout.items, new String[]{"title", "image"}, new int[]{
　　　　　　R.id.title, R.id.image});// 創建 SimpleAdapter
　　　　listview.setAdapter(adapter);// 將適配器與 ListView 關聯
　　　　}
　　}

1.3　擴展練習

（1）編寫 Android 程序,實現通過日期、時間拾取器實現時間設置並通過文本框顯示結果。

（2）編寫 Android 程序,實現單選框選擇「男」「女」並通過文本框顯示結果。

1.4　實驗報告

（1）每人一份實驗報告，統一用學校提供的A4幅面的實驗報告冊書寫或用A4的紙打印。如果打印必須有以下格式的表頭：

實驗課程						
實驗名稱						
實驗時間		學年	學期	週	星期	第　節
學生姓名		學號		班級		
同組姓名		學號		班級		
實驗地點		設備號		指導教師		

（2）實驗內容的主要結果及對結果的分析。
（3）實驗過程中你所遇到的問題的解決辦法。
（4）心得體會、意見和建議。

1.5　實驗成績考核

（1）考勤占10%。
（2）相互協作完成實驗任務占40%。
（3）實驗報告占50%。

實驗二 Andriod 高級 UI 設計

2.1 實驗目的

本次實驗的目的是讓大家熟悉 Android 開發中的高級 UI 設計，包括瞭解和熟悉常用高級控件的使用、消息提示框和對話框等相關內容，通過這些組件，可以開發出更優秀的用戶界面。

2.2 實驗要求

(1) 熟悉和掌握界面高級控件設計和使用。
(2) 掌握進度條的用法。
(3) 掌握 GridView 的用法。
(4) 掌握 AltertDialog 的用法。

2.3 實驗內容

Android 高級控件。
進度條：ProgressBar，ProgressDialog，SeekBar，RatingBar。
選擇器：DatePicker，TimePicker。
菜單：Menu，ContentMenu。
對話框：Dialog，ProgressBar，AltertDialog。

【練習 2.1】運用進度條

在主頁面上設計一個按鈕，在單擊按鈕之後開始線程的週期，在運行的過程中顯示 ProgressBar。

(1) 運行效果如圖 2.1 所示。

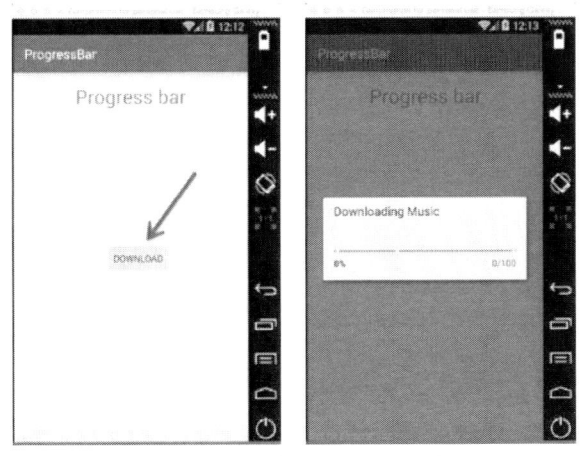

图2.1 进度条练习运行效果图

(2)资源文件佈局如图2.2所示。

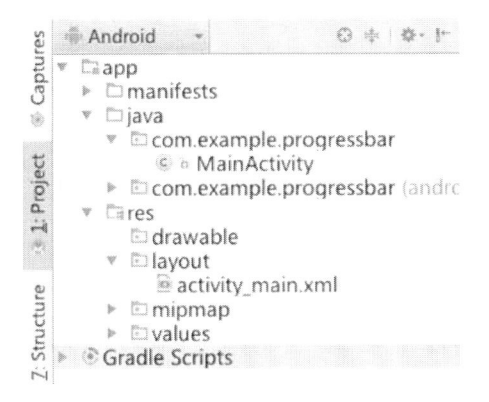

图2.2 进度条练习工程结构图

・activity_main.xml 文件代码如下:

<? xml version = "1.0" encoding = "utf-8" ? >
< RelativeLayout xmlns: android = " http://schemas.android.com/apk/res/android"
　　　xmlns:tools = "http://schemas.android.com/tools"
android:layout_width = "match_parent"
　　　android:layout_height = "match_parent"
android:paddingLeft = "@dimen/activity_horizontal_margin"
　　　android:paddingRight = "@dimen/activity_horizontal_margin"
　　　android:paddingTop = "@dimen/activity_vertical_margin"
　　　android:paddingBottom = "@dimen/activity_vertical_margin" tools:context = ".MainActivity" >

　　　<TextView

```
        android:layout_width="wrap_content"
        android:layout_height="wrap_content"
        android:id="@+id/textView"
        android:layout_alignParentTop="true"
        android:layout_centerHorizontal="true"
        android:textSize="30dp"
        android:text="Progress bar" />

    <Button
        android:layout_width="wrap_content"
        android:layout_height="wrap_content"
        android:text="Download"
        android:onClick="download"
        android:id="@+id/button2"
        android:layout_marginLeft="125dp"
        android:layout_marginStart="125dp"
        android:layout_centerVertical="true" />
</RelativeLayout>
```

· MainActivity.java 文件代碼為：

```java
package com.example.progressbar;

import android.app.ProgressDialog;
import android.os.Bundle;
import android.support.v7.app.AppCompatActivity;
import android.view.View;
import android.widget.Button;

public class MainActivity extends AppCompatActivity {
    Button b1;
    private ProgressDialog progress;
    @Override
    protected void onCreate(Bundle savedInstanceState) {
        super.onCreate(savedInstanceState);
        setContentView(R.layout.activity_main);
        b1 = (Button) findViewById(R.id.button2);
    }
```

```java
public void download(View view){
    progress=new ProgressDialog(this);
    progress.setMessage("Downloading Music");
    progress.setProgressStyle(ProgressDialog.STYLE_HORIZONTAL);
    progress.setIndeterminate(true);
    progress.setProgress(0);
    progress.show();

    final int totalProgressTime = 100;
    final Thread t = new Thread(){
        @Override
        public void run(){
            int jumpTime = 0;

            while(jumpTime < totalProgressTime){
                try{
                    sleep(200);
                    jumpTime += 5;
                    progress.setProgress(jumpTime);
                }
                catch (InterruptedException e){
                    // TODO Auto-generated catch block
                    e.printStackTrace();
                }
            }
        }
    };
    t.start();
}
```

【練習 2.2】運用 GridView

編寫 Android 程序,實現帶預覽的圖片遊覽器。

(1)運行效果如圖 2.3 所示。

图 2.3　GridView 练习运行效果图

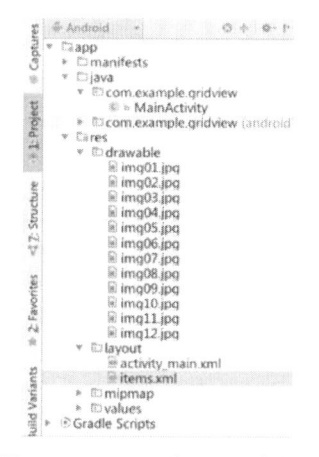

（2）资源文件布局如图 2.4 所示。

图 2.4　GridView 练习工程结构图

・activity_main.xml 文件代码如下：

```
<? xml version = "1.0" encoding = "utf-8" ? >
<LinearLayout xmlns:android = "http://schemas.android.com/apk/res/android"
    android:orientation = "horizontal"
    android:layout_width = "fill_parent"
    android:layout_height = "fill_parent"
    android:id = "@ +id/llayout"
>

<GridView android:id = "@ +id/gridView1"
    android:layout_height = "match_parent"
    android:layout_width = "640px"
```

```xml
        android:layout_marginTop="10px"
        android:horizontalSpacing="3px"
        android:verticalSpacing="3px"
        android:numColumns="4"
        />

<!-- 添加一個圖像切換器 -->
<ImageSwitcher
        android:id="@+id/imageSwitcher1"
        android:padding="20px"
        android:layout_width="match_parent"
        android:layout_height="match_parent"/>
</LinearLayout>
```

· items.xml 文件代碼如下:

```xml
<?xml version="1.0" encoding="utf-8"?>
<LinearLayout
    xmlns:android="http://schemas.android.com/apk/res/android"
    android:orientation="vertical"
    android:gravity="center_horizontal"
    android:layout_width="match_parent"
    android:layout_height="match_parent">
<ImageView
        android:id="@+id/image"
        android:paddingLeft="10px"
        android:paddingTop="20px"
        android:paddingBottom="20px"
        android:adjustViewBounds="true"
        android:maxWidth="300px"
        android:maxHeight="226px"
        android:layout_height="wrap_content"
        android:layout_width="wrap_content"/>
<TextView
        android:layout_width="wrap_content"
        android:layout_height="wrap_content"
        android:layout_gravity="center"
        android:id="@+id/title"
        />
</LinearLayout>
```

・MainActivity.java 文件代碼為：

package com.example.gridview;

import android.app.Activity;
import android.os.Bundle;
import android.view.View;
import android.view.ViewGroup.LayoutParams;
import android.view.animation.AnimationUtils;
import android.widget.AdapterView;
import android.widget.AdapterView.OnItemClickListener;
import android.widget.GridView;
import android.widget.ImageSwitcher;
import android.widget.ImageView;
import android.widget.SimpleAdapter;
import android.widget.ViewSwitcher.ViewFactory;

import java.util.ArrayList;
import java.util.HashMap;
import java.util.List;
import java.util.Map;

public class MainActivity extends Activity {
    private int[] imageId = new int[]{ R.drawable.img01, R.drawable.img02,
        R.drawable.img03, R.drawable.img04, R.drawable.img05,
        R.drawable.img06, R.drawable.img07, R.drawable.img08,
        R.drawable.img09, R.drawable.img10, R.drawable.img11,
        R.drawable.img12, }; // 定義並初始化保存圖片 id 的數組
    final String[] filename = new String[]{ "img01.jpg", "img02.jpg", "img03.jpg", "img04.jpg",
        "img05.jpg", "img06.jpg", "img07.jpg", "img08.jpg", "img09.jpg",
        "img10.jpg", "img11.jpg", "img12.jpg" }; // 定義並初始化保存列表項文字的數組

    private ImageSwitcher imageSwitcher; // 聲明一個圖像切換器對象

```java
@Override
public void onCreate(Bundle savedInstanceState) {
    super.onCreate(savedInstanceState);
    setContentView(R.layout.activity_main);
    /***************** 使用 SimpleAdapter 指定要顯示的內容 *********************/
    List<Map<String, Object>> listItems = new ArrayList<Map<String, Object>>();    // 創建一個 list 集合
    // 通過 for 循環將圖片 id 和列表項文字放到 Map 中, 並添加到 list 集合中
    for (int i = 0; i < imageId.length; i++) {
        Map<String, Object> map = new HashMap<String, Object>();   // 實例化 Map 對象
        map.put("image", imageId[i]);
        map.put("title", filename[i]);
        listItems.add(map);    // 將 map 對象添加到 List 集合中
    }

    final SimpleAdapter adapter = new SimpleAdapter(this, listItems,
            R.layout.items, new String[]{"title", "image"}, new int[]{
            R.id.title, R.id.image});    // 創建 SimpleAdapter

    /***********************************/
    imageSwitcher = (ImageSwitcher) findViewById(R.id.imageSwitcher1);    // 獲取圖像切換器
    // 設置動畫效果
    imageSwitcher.setInAnimation(AnimationUtils.loadAnimation(this,
            android.R.anim.fade_in));    // 設置淡入動畫
    imageSwitcher.setOutAnimation(AnimationUtils.loadAnimation(this,
            android.R.anim.fade_out));    // 設置淡出動畫
    imageSwitcher.setFactory(new ViewFactory() {
```

```
                @Override
                public View makeView(){
                    ImageView imageView = new ImageView(MainActivity.this);
                    // 實例化一個 ImageView 類的對象
                    imageView.setScaleType(ImageView.ScaleType.FIT_CENTER);// 設置保持縱橫比居中縮放圖像
                    imageView.setLayoutParams(new ImageSwitcher.LayoutParams(
                        LayoutParams.WRAP_CONTENT,
                        LayoutParams.WRAP_CONTENT));
                    return imageView;// 返回 imageView 對象
                }
            });
            imageSwitcher.setImageResource(imageId[6]);// 設置默認顯示的圖像
            GridView gridview = (GridView) findViewById(R.id.gridView1);// 獲取 GridView 組件
            gridview.setAdapter(adapter);// 將適配器與 GridView 關聯
            gridview.setOnItemClickListener(new OnItemClickListener(){
                @Override
                public void onItemClick(AdapterView<?> parent, View view, int position,
                    long id){
                    imageSwitcher.setImageResource(imageId[position]);// 顯示選中的圖片
                }
            });
        }
    }
```

【練習 2.3】運用 AltertDialog

編寫 Android 程序，應用 AltertDialog 實現自定義的登錄對話框。

(1) 運行效果如圖 2.5 所示。

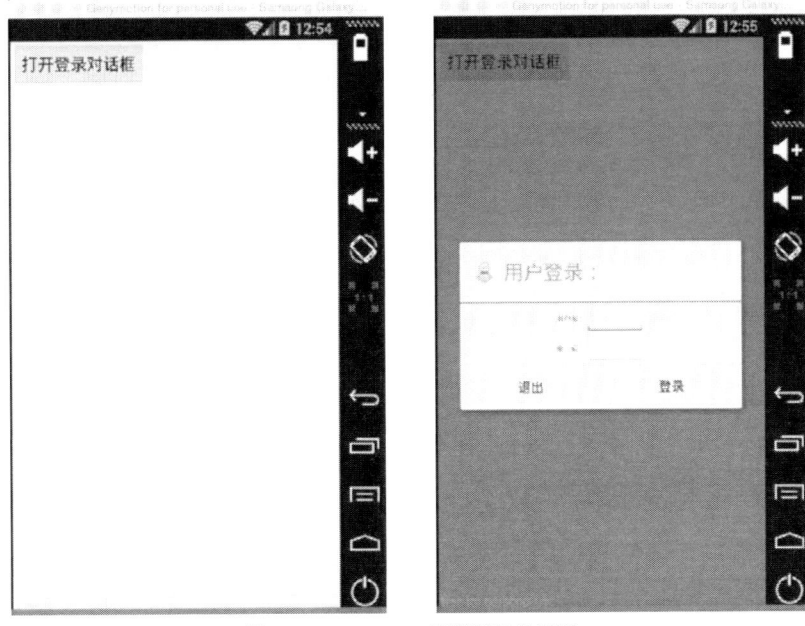

圖 2.5　AlterDialog 練習運行效果圖

(2) 資源文件佈局如圖 2.6 所示。

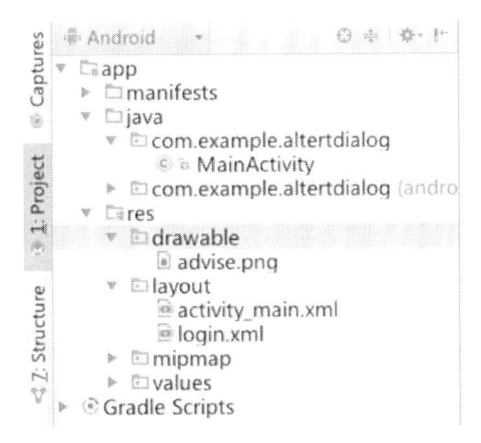

圖 2.6　AlterDialog 練習工程結構圖

・activity_main.xml 文件代碼如下：

```xml
<?xml version="1.0" encoding="utf-8"?>
<LinearLayout xmlns:android="http://schemas.android.com/apk/res/android"
    android:orientation="vertical"
    android:layout_width="fill_parent"
    android:layout_height="fill_parent"
    >
<Button
        android:id="@+id/button1"
        android:layout_width="wrap_content"
        android:layout_height="wrap_content"
        android:text="打開登錄對話框" />
</LinearLayout>
```

· longin.xml 文件代碼如下：

```xml
<?xml version="1.0" encoding="utf-8"?>
<TableLayout android:id="@+id/tableLayout1"
    android:layout_width="fill_parent"
    android:layout_height="fill_parent"
    xmlns:android="http://schemas.android.com/apk/res/android"
    android:gravity="center_vertical"
    android:stretchColumns="0,3"
>
<!-- 第一行 -->
<TableRow android:id="@+id/tableRow1"
            android:layout_width="wrap_content"
            android:layout_height="wrap_content" >
<TextView/>
<TextView android:text="用戶名："
            android:id="@+id/textView1"
            android:layout_width="wrap_content"
            android:textSize="24px"
            android:layout_height="wrap_content"
            />
<EditText android:id="@+id/editText1"
            android:textSize="24px"
            android:layout_width="wrap_content"
            android:layout_height="wrap_content"
android:minWidth="200px"/>
```

```
<TextView />
</TableRow>
<!-- 第二行 -->
<TableRow android:id="@+id/tableRow2"
        android:layout_width="wrap_content"
        android:layout_height="wrap_content">
<TextView/>
<TextView android:text="密    碼："
        android:id="@+id/textView2"
        android:textSize="24px"
        android:layout_width="wrap_content"
        android:layout_height="wrap_content"/>
<EditText android:layout_height="wrap_content"
        android:layout_width="wrap_content"
        android:textSize="24px"
        android:id="@+id/editText2"
        android:inputType="textPassword"/>
<TextView />
</TableRow>
</TableLayout>
```

· MainActivity.java 文件代碼為：

```
package com.example.altertdialog;

import android.app.Activity;
import android.app.AlertDialog;
import android.app.AlertDialog.Builder;
import android.os.Bundle;
import android.view.LayoutInflater;
import android.view.View;
import android.widget.Button;

public class MainActivity extends Activity {
    @Override
    public void onCreate(Bundle savedInstanceState) {
        super.onCreate(savedInstanceState);
```

```
            setContentView(R.layout.activity_main);

            // 自定義的用戶登錄對話框
            Button button1 = (Button) findViewById(R.id.button1); // 獲取佈
局文件中添加的按鈕
            button1.setOnClickListener(new View.OnClickListener() {

                @Override
                public void onClick(View v) {
                    Builder builder = new AlertDialog.Builder(MainActivity.this);
                    builder.setIcon(R.drawable.advise); // 設置對話框的圖標
                    builder.setTitle("用戶登錄:"); // 設置對話框的標題
                    LayoutInflater inflater = getLayoutInflater();
                    View view = inflater.inflate(R.layout.login, null);
                    builder.setView(view);
                    builder.setPositiveButton("登錄", null);
//添加確定按鈕
                    builder.setNegativeButton("退出", null);
//添加取消按鈕
                    builder.create().show(); // 創建對話框並顯示
                }
            });

        }
    }
```

2.4　擴展練習

(1)編寫 Android 程序,通過分別拖動條和星級評分條改變當前文本框數據。
(2)編寫 Android 程序,通過自動完成文本框實現自動文本提示錄入。

2.5　實驗報告

(1)每人一份實驗報告,統一用學校提供的 A4 幅面的實驗報告冊書寫或用 A4 的紙打印。如果打印必須有以下格式的表頭:

實驗課程	
實驗名稱	
實驗時間	學年　　學期　　週　星期　　第　　節
學生姓名	學號　　　　　　班級
同組姓名	學號　　　　　　班級
實驗地點	設備號　　　　　指導教師

（2）實驗內容的主要結果及對結果的分析。
（3）實驗過程中你所遇到的問題的解決辦法。
（4）心得體會、意見和建議。

2.6　實驗成績考核

（1）考勤占 10%。
（2）相互協作完成實驗任務占 40%。
（3）實驗報告占 50%。

實驗三　Intent 與 Activity 的使用

3.1　實驗目的

本次實驗的目的是讓大家熟悉 Intent 和 Activity 的使用。Intent 常用於綁定應用程序組件。Intent 可用於在應用程序 Activity 間啓動、停止和傳輸，並實現添加用戶名、密碼小例程。

3.2　實驗要求

(1) 掌握創建、配置、啓動和關閉 Activity 的方法。
(2) 掌握如何使用 Bundle 在 Activity 之間交換數據。
(3) 掌握 Intent 對象的使用。

3.3　實驗內容

【練習 3.1】從一個 Activity 跳到另一個 Activity

編寫 Android 程序，實現根據輸入的生日判斷星座。
(1) 運行效果如圖 3.1 所示。

圖 3.1　Acitivity1 練習運行效果圖

（2）資源文件佈局為如圖 3.2 所示。

圖 3.2　Acitivity1 練習工程結構圖

- activity_main.xml 文件代碼如下：

```
<?xml version="1.0" encoding="utf-8"?>
<LinearLayout xmlns:android="http://schemas.android.com/apk/res/android"
    android:layout_width="fill_parent"
    android:layout_height="fill_parent"
    android:orientation="vertical" >

<TextView
        android:layout_width="fill_parent"
        android:layout_height="wrap_content"
        android:layout_gravity="center_horizontal"
        android:padding="20dp"
        android:text="計算您的星座" />

<LinearLayout
        android:id="@+id/linearLayout1"
        android:gravity="center_vertical"
        android:layout_width="match_parent"
        android:layout_height="wrap_content" >

<TextView
        android:id="@+id/textView1"
```

```
            android:layout_width="wrap_content"
            android:layout_height="wrap_content"
            android:text="阳历生日:" />

        <EditText
            android:id="@+id/birthday"
            android:minWidth="100dp"
            android:layout_width="wrap_content"
            android:layout_height="wrap_content" >
        </EditText>

        <TextView
            android:id="@+id/textView2"
            android:layout_width="wrap_content"
            android:layout_height="wrap_content"
            android:text="格式:YYYY-MM-DD 例如:2012-01-01" />

    </LinearLayout>

    <Button
        android:id="@+id/button1"
        android:layout_width="wrap_content"
        android:layout_height="wrap_content"
        android:text="确定" />

</LinearLayout>
```

· result.xml 文件代码如下:

```
<?xml version="1.0" encoding="utf-8"?>
<LinearLayout xmlns:android="http://schemas.android.com/apk/res/android"
    android:layout_width="match_parent"
    android:layout_height="match_parent"
    android:orientation="vertical" >

    <TextView
```

```
            android:id="@+id/birthday"
            android:layout_width="wrap_content"
            android:layout_height="wrap_content"
            android:padding="10px"
            android:text="陽歷生日" />

    <TextView
            android:id="@+id/result"
            android:padding="10px"
            android:layout_width="wrap_content"
            android:layout_height="wrap_content"
            android:text="星座" />
</LinearLayout>
```

・MainActivity.java 文件代碼為：

```
package com.example.activity1;

import android.app.Activity;
import android.content.Intent;
import android.os.Bundle;
import android.view.View;
import android.view.View.OnClickListener;
import android.widget.Button;
import android.widget.EditText;
import android.widget.Toast;

public class MainActivity extends Activity {
    @Override
    public void onCreate(Bundle savedInstanceState) {
        super.onCreate(savedInstanceState);
        setContentView(R.layout.activity_main);
        Button button = (Button)findViewById(R.id.button1);
        button.setOnClickListener(new OnClickListener() {

            @Override
            public void onClick(View v) {
```

```
                    Info info=new Info();    //實例化一個保存輸入基本信息
的對象
    if("".equals(((EditText)findViewById(R.id.birthday)).getText().toString
())){
                    Toast.makeText(MainActivity.this,"請輸入您的陽歷生日,
否則不能計算!",Toast.LENGTH_SHORT).show();
                    return;
        }
            String
birthday=((EditText)findViewById(R.id.birthday)).getText().toString();

            info.setBirthday(birthday);   //設置生日
            Bundle bundle=new Bundle();    //實例化一個Bundle對象
            bundle.putSerializable("info",info);   //將輸入的基本信息保
存到Bundle對象中
            Intent intent=new
Intent(MainActivity.this,ResultActivity.class);
            intent.putExtras(bundle);    //將bundle保存到Intent對象中
            startActivity(intent);   //啓動intent對應的Activity
        }
    });
    }
}
```

- Info.java 文件代碼為：

```
package com.example.activity1;

import java.io.Serializable;

public class Info implements Serializable {

    private static final long serialVersionUID = 1L;
    private String birthday="";    //生日
    public String getBirthday() {
        return birthday;
    }
    public void setBirthday(String birthday) {
        this.birthday = birthday;
    }
```

・ResultActivity.java 文件代碼為：

```java
package com.example.activity1;

import android.app.Activity;
import android.content.Intent;
import android.os.Bundle;
import android.widget.TextView;

public class ResultActivity extends Activity {

    @Override
    protected void onCreate(Bundle savedInstanceState) {
        super.onCreate(savedInstanceState);
        setContentView(R.layout.result); // 設置該 Activity 使用的佈局
        TextView birthday = (TextView) findViewById(R.id.birthday); // 獲取顯示生日的文本框
        TextView result = (TextView) findViewById(R.id.result); // 獲取顯示星座的文本框
        Intent intent = getIntent(); // 獲取 Intent 對象
        Bundle bundle = intent.getExtras(); // 獲取傳遞的數據包
        Info info = (Info) bundle.getSerializable("info"); // 獲取一個可序列化的 info 對象
        birthday.setText("您的陽曆生日是" + info.getBirthday()); // 獲取性別並顯示到相應文本框中

        result.setText( query(info.getBirthday())); // 顯示計算後的星座
    }

    /**
     * 功能根據生日查詢星座
     *
     * param month
     * param day
     * return
     */
```

```java
public String query(String birthday){
    int month=0;
    int day=0;
    try{
        month=Integer.parseInt(birthday.substring(5,7));
        day=Integer.parseInt(birthday.substring(8,10));
    }catch(Exception e){
        e.printStackTrace();
    }
    String name = "";// 提示信息
    if(month > 0 && month < 13 && day > 0 && day < 32){ // 如果输入的月和日有效
        if((month == 3 && day > 20) || (month == 4 && day < 21)){
            name = "您是白羊座!";
        }else if((month == 4 && day > 20) || (month == 5 && day < 21)){
            name = "您是金牛座!";
        }else if((month == 5 && day > 20) || (month == 6 && day < 22)){
            name = "您是雙子座!";
        }else if((month == 6 && day > 21) || (month == 7 && day < 23)){
            name = "您是巨蟹座!";
        }else if((month == 7 && day > 22) || (month == 8 && day < 23)){
            name = "您是獅子座!";
        }else if((month == 8 && day > 22) || (month == 9 && day < 23)){
            name = "您是處女座!";
        }else if((month == 9 && day > 22) || (month == 10 && day < 23)){
            name = "您是天平座!";
        }else if((month == 10 && day > 22) || (month == 11 && day < 22)){
            name = "您是天蠍座!";
        }else if((month == 11 && day > 21) || (month == 12 && day < 22)){
            name = "您是射手座!";
        }else if((month == 12 && day > 21) || (month == 1 && day < 20)){
```

```
                    name = "您是摩羯座!";
            } else if ((month == 1 && day > 19) || (month == 2 &&
day < 19)) {
                    name = "您是水牛座!";
            } else if ((month == 2 && day > 18) || (month == 3 &&
day < 21)) {
                    name = "您是雙魚座!";
            }
                    name = month + "月" + day + "日   " + name;
            } else {// 如果輸入的月和日無效
                    name = "您輸入的生日格式不正確或者不是真實
生日!";
            }
                    return name;// 返回星座或提示信息
        }
    }
```

- AndroidManifest.xml 文件代碼如下:

```xml
<?xml version="1.0" encoding="utf-8"?>
<manifest xmlns:android="http://schemas.android.com/apk/res/android"
    package="com.example.activity1">

<application
        android:allowBackup="true"
        android:icon="@mipmap/ic_launcher"
        android:label="@string/app_name"
        android:supportsRtl="true"
        android:theme="@style/AppTheme">
<activity android:name=".MainActivity">
<intent-filter>
<action android:name="android.intent.action.MAIN" />

<category android:name="android.intent.category.LAUNCHER" />
</intent-filter>
</activity>
<activity android:name=".ResultActivity">
</activity>
</application>

</manifest>
```

【練習 3.2】從一個 Activity 跳到另一個 Activity 再返回

編寫 Android 程序，實現帶選擇所在城市的用戶註冊。
(1) 運行效果如圖 3.3 所示。

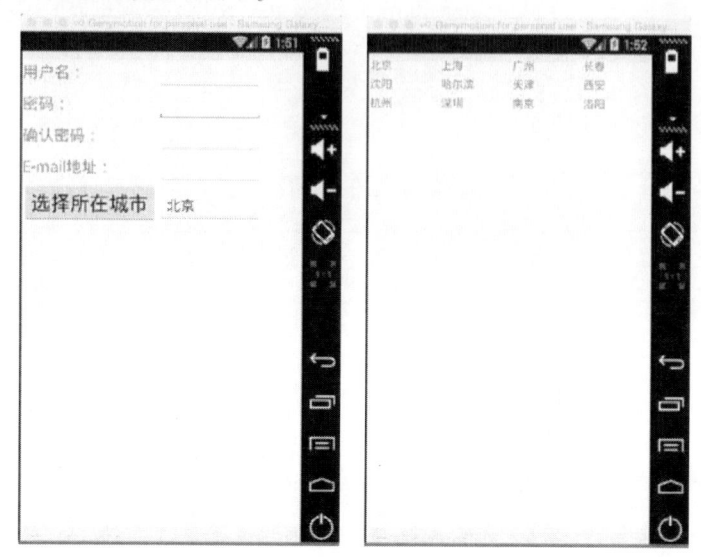

圖 3.3　Acitivity2 練習運行效果圖

(2) 資源文件佈局如圖 3.4 所示。

圖 3.4　Acitivity2 練習工程結構圖

- activity_main.xml 文件代碼如下：

```xml
<?xml version="1.0" encoding="utf-8"?>
<LinearLayout xmlns:android="http://schemas.android.com/apk/res/android"
    android:layout_width="fill_parent"
    android:layout_height="fill_parent"
    android:orientation="horizontal"
    android:paddingTop="20px" >

    <TableLayout
        android:id="@+id/tableLayout1"
        android:layout_width="match_parent"
        android:layout_height="wrap_content" >

        <TableRow
            android:id="@+id/tableRow1"
            android:layout_width="wrap_content"
            android:layout_height="wrap_content" >

            <TextView
                android:id="@+id/textView1"
                android:layout_width="wrap_content"
                android:layout_height="wrap_content"
                android:text="用戶名："
                android:textSize="60px" />
            <EditText
                android:id="@+id/user"
                android:layout_width="wrap_content"
                android:layout_height="wrap_content"
                android:minWidth="400px" />
        </TableRow>

        <TableRow
            android:id="@+id/tableRow2"
            android:layout_width="wrap_content"
            android:layout_height="wrap_content" >

            <TextView
                android:id="@+id/textView2"
```

```xml
            android:layout_width="wrap_content"
            android:layout_height="wrap_content"
            android:text="密碼:"
            android:textSize="60px" />

        <EditText
            android:id="@+id/pwd"
            android:layout_width="wrap_content"
            android:layout_height="wrap_content"
            android:inputType="textPassword" />
    </TableRow>

    <TableRow
        android:id="@+id/tableRow3"
        android:layout_width="wrap_content"
        android:layout_height="wrap_content" >

        <TextView
            android:id="@+id/textView3"
            android:layout_width="wrap_content"
            android:layout_height="wrap_content"
            android:text="確認密碼:"
            android:textSize="60px" />

        <EditText
            android:id="@+id/repwd"
            android:layout_width="wrap_content"
            android:layout_height="wrap_content"
            android:inputType="textPassword" />
    </TableRow>

    <TableRow
        android:id="@+id/tableRow4"
        android:layout_width="wrap_content"
        android:layout_height="wrap_content" >
```

```xml
<TextView
    android:id="@+id/textView3"
    android:layout_width="wrap_content"
    android:layout_height="wrap_content"
    android:text="E-mail 地址:"
    android:textSize="60px" />

<EditText
    android:id="@+id/email"
    android:layout_width="wrap_content"
    android:layout_height="wrap_content" />
    </TableRow>

<TableRow
    android:id="@+id/tableRow5"
    android:layout_width="wrap_content"
    android:layout_height="wrap_content" >

    <Button
        android:id="@+id/button1"
        android:layout_width="wrap_content"
        android:layout_height="wrap_content"
        android:text="選擇所在城市"
        android:textSize="25dp" />

    <EditText
        android:id="@+id/city"
        android:layout_width="wrap_content"
        android:layout_height="wrap_content" />


    </TableRow>
    </TableLayout>

</LinearLayout>
```

・head.xml 文件代碼如下：

```xml
<?xml version="1.0" encoding="utf-8"?>
<LinearLayout xmlns:android="http://schemas.android.com/apk/res/android"
    android:layout_width="match_parent"
    android:layout_height="match_parent"
    android:orientation="vertical" >
    <GridView android:id="@+id/gridView1"
        android:layout_height="match_parent"
        android:layout_width="match_parent"
        android:layout_marginTop="10px"
        android:horizontalSpacing="3px"
        android:verticalSpacing="3px"
        android:numColumns="4"
        />

</LinearLayout>
```

・MainActivity.java 文件代碼為：

```java
package com.example.activity2;

import android.app.Activity;
import android.content.Intent;
import android.os.Bundle;
import android.view.View;
import android.view.View.OnClickListener;
import android.widget.Button;
import android.widget.TextView;

public class MainActivity extends Activity {
    /** Called when the activity is first created. */
    @Override
    public void onCreate(Bundle savedInstanceState) {
        super.onCreate(savedInstanceState);
        setContentView(R.layout.activity_main);
        Button button = (Button)findViewById(R.id.button1);    //獲取選擇頭像按鈕
        button.setOnClickListener(new OnClickListener() {
```

```java
        @Override
        public void onClick(View v){
            Intent intent=new Intent(MainActivity.this,HeadActivity.class);
            startActivityForResult(intent, 0x11);    //啓動 intent 對應的 Activity
        }
    });
}

@Override
protected void onActivityResult(int requestCode, int resultCode, Intent data){
    super.onActivityResult(requestCode, resultCode, data);
    if(requestCode==0x11 && resultCode==0x11){  //判斷是否為待處理的結果
        Bundle bundle=data.getExtras();           //獲取傳遞的數據包
        String city=bundle.getString("city");    //獲取選擇的頭像 ID
        TextView tv=(TextView)findViewById(R.id.city);  //獲取佈局文件中添加的 ImageView 組件
        tv.setText(city);
    }
}
```

‧HeadActivity.java 文件代碼為：

```java
package com.example.activity2;

import android.app.Activity;
import android.content.Intent;
import android.os.Bundle;
import android.view.View;
import android.view.ViewGroup;
import android.widget.AdapterView;
import android.widget.AdapterView.OnItemClickListener;
import android.widget.BaseAdapter;
import android.widget.GridView;
```

```
import android.widget.TextView;
public class HeadActivity extends Activity {

    public String[] city = new String[] { "北京","上海","廣州","長春","沈陽","哈爾濱","天津","西安","杭州","深圳","南京","洛陽" };
                    // 定義並初始化保存頭像 id 的數組
    @Override
    protected void onCreate( Bundle savedInstanceState) {
        super.onCreate( savedInstanceState);
        setContentView( R.layout.head);          //設置該 Activity 使用的佈局
        GridView gridview = (GridView) findViewById( R.id.gridView1);
// 獲取 GridView 組件
        BaseAdapter adapter = new BaseAdapter() {
            @Override
            public View getView( int position, View convertView, ViewGroup parent) {

                TextView tv;                          //聲明 ImageView 的對象
                if( convertView = = null) {
                    tv = new TextView( HeadActivity.this);        //實例化 ImageView 的對象
//              /************設置圖像的寬度和高度**************/
//                  imageview.setAdjustViewBounds( true);
//                  imageview.setMaxWidth( 158);
//                  imageview.setMaxHeight( 150);
//              /*******************************/
                    tv.setPadding( 5, 5, 5, 5);              //設置 ImageView 的內邊距
                } else {
                    tv = ( TextView) convertView;
                }
                tv.setText( city[ position]);              //為 ImageView 設置要顯示的圖片
                return tv; //返回 ImageView
            }
            /*
```

```java
         * 功能:獲得當前選項的 ID
         */
        @Override
        public long getItemId(int position) {
            return position;
        }
        /*
         * 功能:獲得當前選項
         */
        @Override
        public Object getItem(int position) {
            return position;
        }
        /*
         * 獲得數量
         */
        @Override
        public int getCount() {
            return city.length;
        }
    };

    gridview.setAdapter(adapter);                              // 將適配器與 GridView 關聯
    gridview.setOnItemClickListener(new OnItemClickListener() {
        @Override
        public void onItemClick(AdapterView<?> parent, View view, int position, long id) {
            Intent intent = getIntent();  //獲取 Intent 對象
            Bundle bundle = new Bundle();       //實例化要傳遞的數據包
            bundle.putString("city", city[position]);// 顯示選中的圖片
            intent.putExtras(bundle);   //將數據包保存到 intent 中
            setResult(0x11, intent);        //設置返回的結果碼,並返回調用該 Activity 的 Activity
            finish();      //關閉當前 Activity
        }
    });

}
```

· AndroidManifest.xml 文件代碼如下：

```xml
<?xml version="1.0" encoding="utf-8"?>
<manifest xmlns:android="http://schemas.android.com/apk/res/android"
    package="com.example.activity2" >

    <application
        android:allowBackup="true"
        android:icon="@mipmap/ic_launcher"
        android:label="@string/app_name"
        android:supportsRtl="true"
        android:theme="@style/AppTheme" >
        <activity android:name=".MainActivity" >
            <intent-filter>
                <action android:name="android.intent.action.MAIN" />

                <category android:name="android.intent.category.LAUNCHER" />
            </intent-filter>
        </activity>
        <activity android:name=".HeadActivity" >
        </activity>
    </application>

</manifest>
```

3.4 擴展練習

（1）編寫 Android 程序，實現查看全部 Activity 生命週期回調函數，通過 toast 方法返回結果。

（2）編寫 Android 程序，實現使用 Intent 撥打電話。

3.5 實驗報告

（1）每人一份實驗報告，統一用學校提供的 A4 幅面的實驗報告冊書寫或用 A4 的紙打印。如果打印必須有以下格式的表頭：

實驗課程					
實驗名稱					
實驗時間	學年　　學期　　週　星期　　第　　節					
學生姓名		學號		班級	
同組姓名		學號		班級	
實驗地點		設備號		指導教師	

(2)實驗內容的主要結果及對結果的分析。
(3)實驗過程中你所遇到的問題的解決辦法。
(4)心得體會、意見和建議。

3.6　實驗成績考核

(1)考勤占 10%。
(2)相互協作完成實驗任務占 40%。
(3)實驗報告占 50%。

實驗四　Android 資源訪問

4.1　實驗目的

本次實驗的目的是讓大家熟悉 Android 中的資源，資源指的是代碼中使用的外部文件，這些文件作為應用程序的一部分，被編譯到應用程序中。

4.2　實驗要求

(1)掌握字符串資源、顏色資源和尺寸資源文件的定義和使用。
(2)掌握如何通過菜單資源定義上下文菜單和選項菜單。

4.3　實驗內容

【練習 4.1】為 ImageView 更換圖片

實現本程序前需要準備兩張圖片，分別存放在「/res/drawable/」文件夾與手機文件系統的另一個文件夾中。在本程序中兩張圖片文件的路徑分別為/res/drawable/img01.jpg 與/data/data/com.example.resource1/img02.jpg。程序中會先將 drawable 目錄裡的圖片文件顯示在 ImageView 中，再設計一個 Button，當用戶單擊 Button 後，將 ImageView 裡的圖片換成另一張存在手機文件系統裡的圖形文件。

(1)運行效果如圖 4.1 所示。

图 4.1　更改 ImageView 资源练习运行效果图

（2）将 img02.jpg 导入手机文件系统。

选择 Android Studio 菜单栏「Tools」→「Android」→「Android Device Monitor」，如图 4.2 所示。

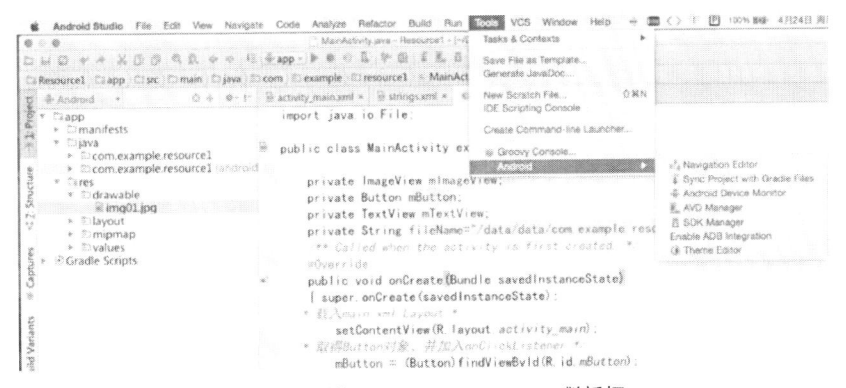

图 4.2　打开 Android Device Monitor 对话框

在弹出来的 Android Device Monitor 窗体中，点击「File Explorer」，找到将要导入图片的路径:/data/data/com.example.resource1，然后点击「　　」图标导入 img02.jpg 文件，最后效果如图 4.3 所示。

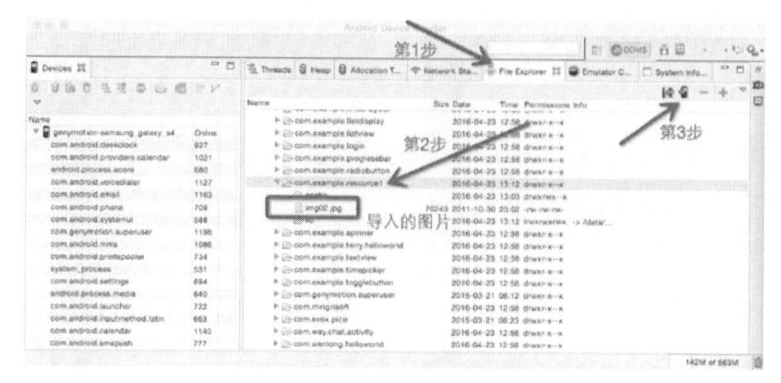

圖 4.3　將圖片導入手機文件系統

（3）資源文件佈局如圖 4.4 所示。

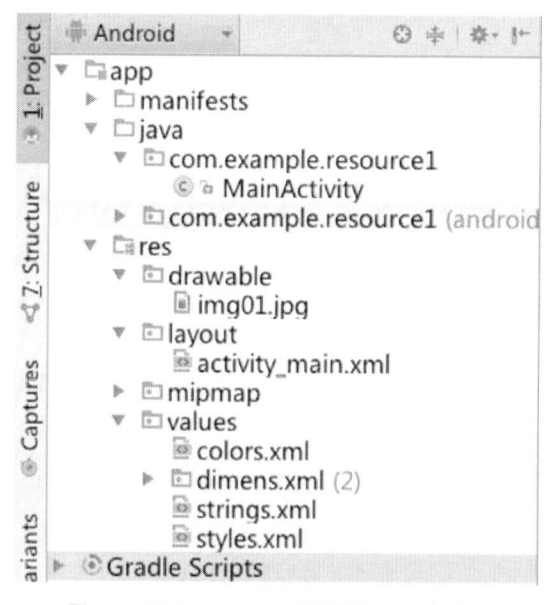

圖 4.4　更改 ImageView 資源練習工程結構圖

・activity_main.xml 文件代碼如下：

<?xml version = "1.0" encoding = "utf-8" ? >
<RelativeLayout xmlns：android = "http://schemas.android.com/apk/res/android"
　　xmlns：tools = " http://schemas.android.com/tools"　android：layout_width = "match_parent"
　　android：layout_height = " match_parent"　android：paddingLeft = " @ dimen/activity_horizontal_margin"

```xml
        android:paddingRight = "@dimen/activity_horizontal_margin"
        android:paddingTop = "@dimen/activity_vertical_margin"
        android:paddingBottom = "@dimen/activity_vertical_margin" tools:context = ".MainActivity" >

    <TextView
        android:id = "@+id/mTextView"
        android:layout_width = "fill_parent"
        android:layout_height = "wrap_content"
        android:text = "@string/str_text"
        android:layout_x = "30px"
        android:layout_y = "30px"
        />
    <ImageView
        android:id = "@+id/mImageView"
        android:layout_width = "500px"
        android:layout_height = "356px"
        android:src = "@drawable/img01"
        android:layout_x = "30px"
        android:layout_y = "62px"
        android:layout_marginTop = "87dp"
        android:layout_below = "@+id/mTextView"
        android:layout_centerHorizontal = "true" />
    <Button
        android:id = "@+id/mButton"
        android:layout_width = "155px"
        android:layout_height = "wrap_content"
        android:text = "@string/str_button"
        android:textSize = "18sp"
        android:layout_x = "80px"
        android:layout_y = "302px"
        android:layout_centerVertical = "true"
        android:layout_alignStart = "@+id/mImageView"
        android:layout_alignEnd = "@+id/mImageView" />
</RelativeLayout>
```

· strings.xml 文件代碼如下：

```xml
<resources>
    <string name="app_name">Resource1</string>
    <string name="str_button">更換圖片</string>
    <string name="str_text">res/drawable/img01.jpg</string>
</resources>
```

· MainActivity.java 文件代碼為：

```java
package com.example.resource1;

import android.app.Activity;
import android.graphics.Bitmap;
import android.graphics.BitmapFactory;
import android.os.Bundle;
import android.view.View;
import android.widget.Button;
import android.widget.ImageView;
import android.widget.TextView;

import java.io.File;

public class MainActivity extends Activity {

    private ImageView mImageView;
    private Button mButton;
    private TextView mTextView;
    private String fileName = "/data/data/com.example.resource1/img02.jpg";
    /** Called when the activity is first created. */
    @Override
    public void onCreate(Bundle savedInstanceState)
    {   super.onCreate(savedInstanceState);
    /* 載入 main.xml Layout */
        setContentView(R.layout.activity_main);
    /* 取得 Button 對象，並加入 onClickListener */
        mButton = (Button)findViewById(R.id.mButton);
        mButton.setOnClickListener(new Button.OnClickListener() {
            public void onClick(View v) {
```

```
        /* 取得對象 */
            mImageView = (ImageView) findViewById(R.id.mImageView);
            mTextView = (TextView) findViewById(R.id.mTextView);
        /* 檢查文件是否存在 */
            File f = new File(fileName);
            if (f.exists()) {
        /* 產生 Bitmap 對象, 並放入 mImageView 中 */
                Bitmap bm = BitmapFactory.decodeFile(fileName);
                mImageView.setImageBitmap(bm);
                mTextView.setText(fileName);
            } else {
                mTextView.setText("文件不存在");
            }
        }
    });
}
```

【練習 4.2】運用上下文菜單

編寫 Android 項目,實現帶子菜單的上下文菜單。
(1)運行效果如圖 4.5 所示。

圖 4.5　上下文菜單練習運行效果圖

(2)資源文件佈局如圖4.6所示。

圖4.6　上下文菜單練習工程結構圖

(3)在res目錄下創建一個menu目錄，並在該目錄中創建一個名稱為contextmenu.xml的菜單資源文件。右鍵單擊res目錄選擇「New」→「Directory」創建menu目錄；右鍵單擊menu目錄選擇「New」→「Menu resource file」創建contextmenu.xml的菜單資源文件。創建方式如圖4.7所示。

圖4.7　創建菜單資源文件

- contextmenu.xml 文件代碼如下：

```xml
<?xml version="1.0" encoding="utf-8"?>
<menu xmlns:android="http://schemas.android.com/apk/res/android">

    <item
        android:id="@+id/color1"
        android:title="紅色">
    </item>
    <item
        android:id="@+id/color2"
        android:title="綠色">
    </item>
    <item
        android:id="@+id/color3"
        android:title="藍色">
    </item>
    <item
        android:id="@+id/color4"
        android:title="橙色">
    </item>
    <item
        android:id="@+id/color5"
        android:title="恢復默認">
    </item>
    <item
        android:id="@+id/item2"
        android:alphabeticShortcut="e"
        android:title="其他顏色">
        <menu>
            <group
                android:id="@+id/other">
                <item
                    android:id="@+id/other1"
                    android:title="橄欖色">
                </item>
                <item
                    android:id="@+id/other2"
                    android:title="水綠色">
                </item>
            </group>
        </menu>
    </item>

</menu>
```

· activity_main.xml 文件代碼如下：

```xml
<?xml version="1.0" encoding="utf-8"?>
<LinearLayout xmlns:android="http://schemas.android.com/apk/res/android"
    android:layout_width="fill_parent"
    android:layout_height="fill_parent"
    android:padding="5px"
    android:orientation="vertical" >
    <TextView
        android:id="@+id/show"
        android:textSize="60px"
        android:layout_width="match_parent"
        android:layout_height="wrap_content"
        android:text="打開菜單..." />

</LinearLayout>
```

· MainActivity.java 文件代碼為：

```java
package com.example.menu;

import android.app.Activity;
import android.graphics.Color;
import android.os.Bundle;
import android.view.ContextMenu;
import android.view.ContextMenu.ContextMenuInfo;
import android.view.MenuInflater;
import android.view.MenuItem;
import android.view.View;
import android.widget.TextView;

public class MainActivity extends Activity {
    private TextView tv;
    @Override
    public void onCreate(Bundle savedInstanceState) {
        super.onCreate(savedInstanceState);
        setContentView(R.layout.activity_main);
        tv = (TextView)findViewById(R.id.show);
```

```
        registerForContextMenu(tv);          //為文本框註冊上下文菜單
    }
    //創建上下文菜單
    /******************************************
    ****************************/
    @Override
    public void onCreateContextMenu(ContextMenu menu, View v,
                        ContextMenuInfo menuInfo){
        MenuInflater inflator = new MenuInflater(this);    //實例化一個MenuInflater對象
        inflator.inflate(R.menu.contextmenu, menu);     //解析菜單文件
        menu.setHeaderIcon(R.drawable.ic_launcher);        //為菜單頭設置圖標
        menu.setHeaderTitle("請選擇文字顏色:");             //為菜單頭設置標題
    }

    @Override
    public boolean onContextItemSelected(MenuItem item){
        if(item.getGroupId() == R.id.other){            //判斷是否選擇了參數設置菜單組
            if(item.getItemId() == R.id.other1){                //當菜單項已經被選中
                tv.setTextColor(Color.rgb(118, 146, 60));
            }else if(item.getItemId() == R.id.other2){
                tv.setTextColor(Color.rgb(0, 255, 255));
            }
        }else{
            switch(item.getItemId()){
                case R.id.color1:     //當選擇紅顏色時
                    tv.setTextColor(Color.rgb(255, 0, 0));
                    break;
                case R.id.color2:     //當選擇綠顏色時
                    tv.setTextColor(Color.rgb(0, 255, 0));
                    break;
                case R.id.color3:     //當選擇藍顏色時
```

```
                    tv.setTextColor(Color.rgb(0, 0, 255));
                break;
            case R.id.color4:        //當選擇橙色時
                tv.setTextColor(Color.rgb(255, 180, 0));
                break;
            default:
                tv.setTextColor(Color.rgb(255, 255, 255));
        }
    }
    return true;
}
```

4.4 擴展練習

編寫 Android 程序,實現帶子菜單的選項菜單。

4.5 實驗報告

(1)每人一份實驗報告,統一用學校提供的 A4 幅面的實驗報告冊書寫或用 A4 的紙打印。如果打印必須有以下格式的表頭:

實驗課程							
實驗名稱							
實驗時間		學年	學期	週	星期	第	節
學生姓名		學號		班級			
同組姓名		學號		班級			
實驗地點		設備號		指導教師			

(2)實驗內容的主要結果及對結果的分析。
(3)實驗過程中你所遇到的問題的解決辦法。
(4)心得體會、意見和建議。

4.6 實驗成績考核

(1) 考勤占 10%。
(2) 相互協作完成實驗任務占 40%。
(3) 實驗報告占 50%。

實驗五　圖形圖像與多媒體

5.1　實驗目的

在屏幕繪制各種圖形，瞭解 Android 中圖形圖像處理技術和多媒體技術。

5.2　實驗要求

(1)瞭解在屏幕繪圖的方法。
(2)瞭解 Android 上圖形圖像處理技術的使用。
(3)瞭解 Android 上多媒體技術的使用。

5.3　實驗內容

【練習 5.1】探照燈效果

編寫 Android 項目，實現探照燈效果。
(1)運行效果如圖 5.1 所示。

圖 5.1　探照燈練習運行效果圖

(2)資源文件佈局如圖 5.2 所示。

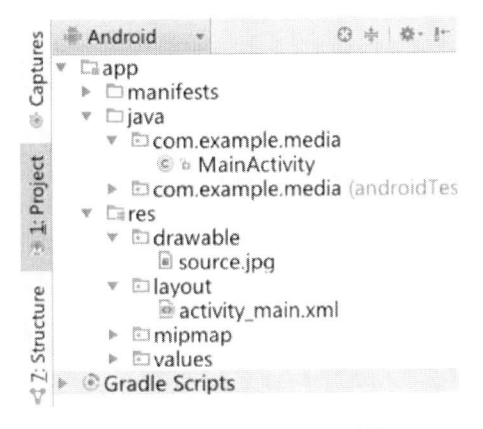

圖 5.2　探照燈練習工程結構圖

・activity_main.xml 文件代碼如下：

```xml
<?xml version="1.0" encoding="utf-8"?>
<FrameLayout xmlns:android="http://schemas.android.com/apk/res/android"
    android:id="@+id/frameLayout1"
    android:layout_width="fill_parent"
    android:layout_height="fill_parent"
    android:orientation="vertical" >

</FrameLayout>
```

・MainActivity.java 文件代碼為：

```
package com.example.media;

import android.app.Activity;
import android.content.Context;
import android.graphics.Bitmap;
import android.graphics.BitmapFactory;
import android.graphics.BitmapShader;
import android.graphics.Canvas;
import android.graphics.Matrix;
import android.graphics.Paint;
import android.graphics.Shader.TileMode;
import android.graphics.drawable.ShapeDrawable;
import android.graphics.drawable.shapes.OvalShape;
```

```java
import android.os.Bundle;
import android.view.MotionEvent;
import android.view.View;
import android.widget.FrameLayout;

public class MainActivity extends Activity {
    @Override
    public void onCreate(Bundle savedInstanceState) {
        super.onCreate(savedInstanceState);
        setContentView(R.layout.activity_main);
        FrameLayout ll = (FrameLayout) findViewById(R.id.frameLayout1); // 獲取佈局文件中的幀佈局管理器
        ll.addView(new MyView(this)); // 將自定義視圖添加到幀佈局管理器中
    }

    public class MyView extends View {
        private Bitmap bitmap; // 源圖像,也就是背景圖像
        private ShapeDrawable drawable;

        private final int RADIUS = 200; // 探照燈的半徑

        private Matrix matrix = new Matrix();

        public MyView(Context context) {
            super(context);
            Bitmap bitmap_source = BitmapFactory.decodeResource(getResources(),
                    R.drawable.source);     //獲取要顯示的源圖像
            bitmap = bitmap_source;
            BitmapShader shader = new BitmapShader(Bitmap.createScaledBitmap(
                    bitmap_source, bitmap_source.getWidth(),
                    bitmap_source.getHeight(), true), TileMode.CLAMP,
                    TileMode.CLAMP);     //創建 BitmapShader 對象
```

```
            // 圓形的 drawable
            drawable = new ShapeDrawable(new OvalShape());
            drawable.getPaint().setShader(shader);
            drawable.setBounds(0, 0, RADIUS * 2, RADIUS * 2); // 設置圓
的外切矩形
        }

        @Override
        protected void onDraw(Canvas canvas) {
            super.onDraw(canvas);
            Paint p=new Paint();
            p.setAlpha(50);
            canvas.drawBitmap(bitmap, 0,0, p); // 繪制背景圖像
            drawable.draw(canvas); // 繪制探照燈照射的圖像
        }

        @Override
        public boolean onTouchEvent(MotionEvent event) {
            final int x = (int) event.getX(); // 獲取當前觸摸點的 X 軸坐標
            final int y = (int) event.getY(); // 獲取當前觸摸點的 Y 軸坐標
            matrix.setTranslate(RADIUS - x , RADIUS - y); // 平移到繪制
shader 的起始位置
            drawable.getPaint().getShader().setLocalMatrix(matrix);
            drawable.setBounds(x - RADIUS, y - RADIUS, x + RADIUS, y +
RADIUS); // 設置圓的外切矩形
            invalidate(); // 重繪畫布
            return true;
        }
    }
}
```

【練習 5.2】實現視頻播放器

編寫 Android 項目,使用 VideoView 實現的視頻播放器。

(1)運行效果如圖 5.3 所示。

圖 5.3　視頻播放器練習運行效果圖

（2）資源文件佈局如圖 5.4 所示。

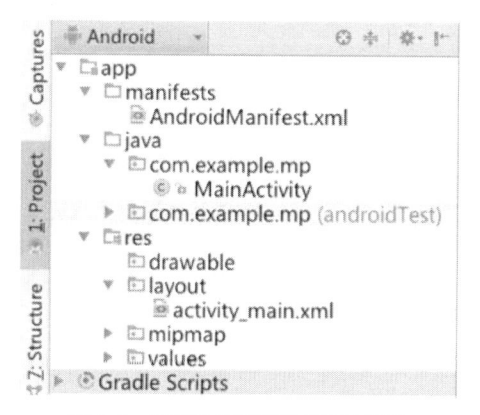

圖 5.4　視頻播放器練習工程結構圖

・activity_main.xml 文件代碼如下：

```
<? xml version = "1.0" encoding = "utf-8" ? >
<RelativeLayout xmlns：android = " http：//schemas.android.com/apk/res/android"
    xmlns：tools = " http：//schemas. android. com/tools "  android：layout_width = " match_parent"
    android：layout_height = " match_parent "  android：paddingLeft = " @ dimen/activity_horizontal_margin"
    android：paddingRight = " @ dimen/activity_horizontal_margin"
    android：paddingTop = " @ dimen/activity_vertical_margin"
    android：paddingBottom = " @ dimen/activity_vertical_margin"  tools：context = " .MainActivity" >
```

```xml
<VideoView
    android:id="@+id/VideoView01"
    android:layout_width="fill_parent"
    android:layout_height="fill_parent"
    android:layout_below="@+id/LoadButton" />
<Button android:id="@+id/LoadButton"
    android:layout_width="wrap_content"
    android:layout_height="wrap_content"
    android:text="装载"
    android:layout_x="30px"
    android:layout_y="300px"
    android:textSize="40dp" />
<Button android:id="@+id/PlayButton"
    android:layout_width="wrap_content"
    android:layout_height="wrap_content"
    android:text="播放"
    android:layout_x="120px"
    android:layout_y="300px"
    android:textSize="40dp"
    android:layout_alignParentTop="true"
    android:layout_toEndOf="@+id/LoadButton" />
<Button android:id="@+id/PauseButton"
    android:layout_width="wrap_content"
    android:layout_height="wrap_content"
    android:text="暂停"
    android:layout_x="210px"
    android:layout_y="300px"
    android:textSize="40dp"
    android:layout_alignParentTop="true"
    android:layout_toEndOf="@+id/PlayButton" />
</RelativeLayout>
```

· MainActivity.java 文件代碼為：

```java
package com.example.mp;

import android.app.Activity;
import android.content.pm.ActivityInfo;
```

```java
import android.os.Bundle;
import android.util.Log;
import android.view.View;
import android.view.View.OnClickListener;
import android.view.Window;
import android.view.WindowManager;
import android.widget.Button;
import android.widget.MediaController;
import android.widget.VideoView;

public class MainActivity extends Activity {
    private static final String TAG = "VideoView";
    @Override
    protected void onCreate(Bundle savedInstanceState) {
        super.onCreate(savedInstanceState);

        //不要標題
        requestWindowFeature(Window.FEATURE_NO_TITLE);
        //設置成全屏模式
        getWindow().setFlags(WindowManager.LayoutParams.FLAG_FULLSCREEN,
                WindowManager.LayoutParams.FLAG_FULLSCREEN);
        //強制為橫屏
        setRequestedOrientation(ActivityInfo.SCREEN_ORIENTATION_LANDSCAPE);

        setContentView(R.layout.activity_main);



        final VideoView videoView = (VideoView) findViewById(R.id.VideoView01);


        Button PauseButton = (Button) this.findViewById(R.id.PauseButton);
        Button LoadButton = (Button) this.findViewById(R.id.LoadButton);
        Button PlayButton = (Button) this.findViewById(R.id.PlayButton);
```

```java
        LoadButton.setOnClickListener( new OnClickListener( )
        {
            public void onClick( View arg0)
            {

                videoView.setVideoPath( "/mnt/sdcard/apple.mp4" );
                videoView.setMediaController( new MediaController( MainActivity.this));
                videoView.requestFocus( );
            }
        });

        PlayButton.setOnClickListener( new OnClickListener( )
        {
            public void onClick( View arg0)
            {

                Log.v( TAG, "start" );
                videoView.start( );
                Log.v( TAG, "start OK" );
            }
        });

        PauseButton.setOnClickListener( new OnClickListener( )
        {
            public void onClick( View arg0)
            {
                videoView.pause( );
            }
        });
    }
}
```

· AndroidManifest.xml 文件代碼為:

```xml
<? xml version="1.0" encoding="utf-8" ? >
<manifest xmlns:android="http://schemas.android.com/apk/res/android"
    package="com.example.mp" >
```

```xml
<application
    android:allowBackup="true"
    android:icon="@mipmap/ic_launcher"
    android:label="@string/app_name"
    android:supportsRtl="true"
    android:theme="@style/AppTheme" >
    <activity android:name=".MainActivity" >
        <intent-filter>
            <action android:name="android.intent.action.MAIN" />

            <category android:name="android.intent.category.LAUNCHER" />
        </intent-filter>
    </activity>
</application>
    <uses-permission android:name="android.permission.WRITE_EXTERNAL_STORAGE"/>
</manifest>
```

（3）將 apple.mp4 視頻文件導入手機文件系統。

選擇 Android Studio 菜單欄「Tools」→「Android」→「Android Device Monitor」，如圖 5.5 所示。

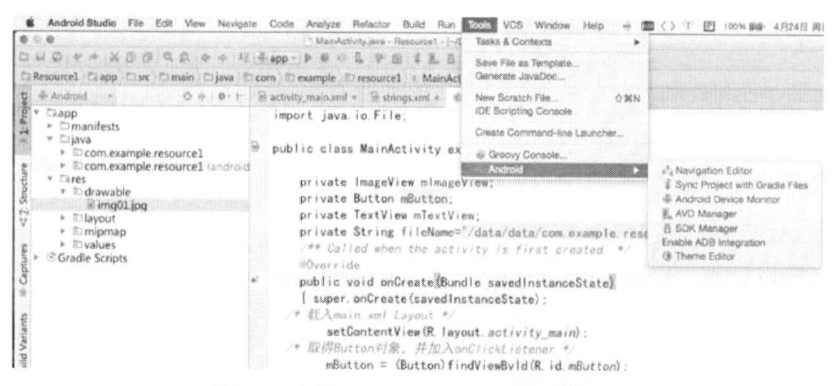

圖 5.5　打開 Android Device Monitor 對話框

在彈出來的 Android Device Monitor 窗體中，點擊「File Explorer」，找到將要導入視頻的路徑：/mnt/sdcard/，然後點擊「　　」圖標導入 apple.mp4 文件，最後效果如圖 5.6 所示。

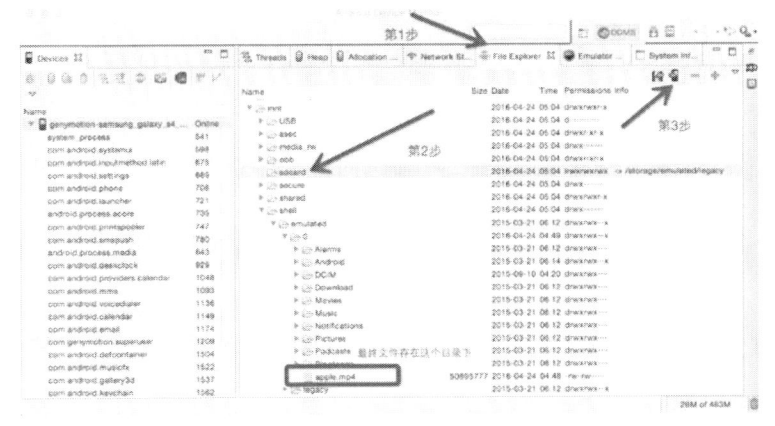

圖 5.6　將視頻導入手機文件系統

5.4　擴展練習

（1）編寫 Android 程序，實現閃爍的星星動畫。
（2）編寫 Android 程序，實現 mp3 音樂播放器。

5.5　實驗報告

（1）每人一份實驗報告，統一用學校提供的 A4 幅面的實驗報告冊書寫或用 A4 的紙打印。如果打印必須有以下格式的表頭：

實驗課程						
實驗名稱						
實驗時間	學年	學期	週	星期	第	節
學生姓名		學號		班級		
同組姓名		學號		班級		
實驗地點		設備號		指導教師		

（2）實驗內容的主要結果及對結果的分析。
（3）實驗過程中你所遇到的問題的解決辦法。
（4）心得體會、意見和建議。

5.6　實驗成績考核

（1）考勤占 10%。
（2）相互協作完成實驗任務占 40%。
（3）實驗報告占 50%。

實驗六 Android 的網路編程基礎

6.1 實驗目的

本次實驗的目的是讓大家熟悉 Android 開發中的如何獲取天氣預報,包括瞭解和熟悉 WebView,WebService 使用,以及網路編程事件處理等內容。

6.2 實驗要求

(1)熟悉和掌握 WebView 使用。
(2)瞭解 Android 的網路編程。
(3)熟悉和掌握 WebService 使用。

6.3 實驗內容

【練習 6.1】基於 WebView 獲取天氣預報

(1)運行效果如圖 6.1 所示。

圖 6.1 WebView 練習運行效果圖

(2)資源文件佈局如圖 6.2 所示。

圖 6.2　WebView 練習工程結構圖

・activity_main.xml 文件代碼如下：

```xml
<?xml version="1.0" encoding="utf-8"?>
<LinearLayout xmlns:android="http://schemas.android.com/apk/res/android"
    android:orientation="vertical"
    android:gravity="center_horizontal"
    android:layout_width="fill_parent"
    android:layout_height="fill_parent"
    >
    <LinearLayout
        android:orientation="horizontal"
        android:layout_width="wrap_content"
        android:layout_height="wrap_content"
        >

        <Button
            android:id="@+id/bj"
            android:layout_width="wrap_content"
            android:layout_height="wrap_content"
            android:text="@string/bj"
            android:textSize="30dp" />

        <Button
            android:id="@+id/sh"
            android:layout_width="wrap_content"
            android:layout_height="wrap_content"
            android:text="@string/sh"
            android:textSize="30dp" />
```

```xml
    <Button
        android:id="@+id/heb"
        android:layout_width="wrap_content"
        android:layout_height="wrap_content"
        android:text="@string/heb"
        android:textSize="30dp" />

</LinearLayout>
<LinearLayout
    android:orientation="horizontal"
    android:layout_width="wrap_content"
    android:layout_height="wrap_content"
    >
    <Button
        android:id="@+id/gz"
        android:layout_width="wrap_content"
        android:layout_height="wrap_content"
        android:text="@string/gz"
        android:textSize="30dp" />

    <Button
        android:id="@+id/cc"
        android:layout_width="wrap_content"
        android:layout_height="wrap_content"
        android:text="@string/cc"
        android:textSize="30dp" />

    <Button
        android:id="@+id/sy"
        android:layout_width="wrap_content"
        android:layout_height="wrap_content"
        android:text="@string/sy"
        android:textSize="30dp"
        android:layout_gravity="right" />
</LinearLayout>
<WebView android:id="@+id/webView1"
    android:layout_width="match_parent"
```

```
            android:layout_height="0dip"
            android:focusable="false"
            android:layout_weight="1"
            />
```

</LinearLayout>

· strings.xml 文件代碼如下:

```
<resources>
<string name="app_name">WebView</string>
<string name="go">GO</string>
<string name="bj">北京</string>
<string name="sh">上海</string>
<string name="gz">廣州</string>
<string name="heb">哈爾濱</string>
<string name="cc">長春</string>
<string name="sy">瀋陽</string>
</resources>
```

· MainActivity.java 文件代碼為:

```java
package com.example.webview;

import android.app.Activity;
import android.os.Bundle;
import android.view.View;
import android.view.View.OnClickListener;
import android.webkit.WebChromeClient;
import android.webkit.WebView;
import android.webkit.WebViewClient;
import android.widget.Button;

public class MainActivity extends Activity implements OnClickListener {
    private WebView webView;    //聲明 WebView 組件的對象

    @Override
    protected void onCreate(Bundle savedInstanceState) {
        super.onCreate(savedInstanceState);
```

```java
        setContentView(R.layout.activity_main);
        webView=(WebView)findViewById(R.id.webView1);        //獲取 WebView 組件
        webView.getSettings().setJavaScriptEnabled(true);     //設置 JavaScript 可用
        webView.setWebChromeClient(new WebChromeClient());    //處理 JavaScript 對話框
        webView.setWebViewClient(new WebViewClient());        //處理各種通知和請求事件,如果不使用該句代碼,將使用內置瀏覽器訪問網頁
        webView.loadUrl("http://m.weather.com.cn/mweather/"); //設置默認顯示的天氣預報信息
        webView.setInitialScale(57*4);  //放網頁內容放大 4 倍
        Button bj=(Button)findViewById(R.id.bj);       //獲取佈局管理器中添加的「北京」按鈕
        bj.setOnClickListener(this);
        Button sh=(Button)findViewById(R.id.sh);       //獲取佈局管理器中添加的「上海」按鈕
        sh.setOnClickListener(this);
        Button heb=(Button)findViewById(R.id.heb);     //獲取佈局管理器中添加的「哈爾濱」按鈕
        heb.setOnClickListener(this);
        Button cc=(Button)findViewById(R.id.cc);       //獲取佈局管理器中添加的「長春」按鈕
        cc.setOnClickListener(this);
        Button sy=(Button)findViewById(R.id.sy);       //獲取佈局管理器中添加的「沈陽」按鈕
        sy.setOnClickListener(this);
        Button gz=(Button)findViewById(R.id.gz);       //獲取佈局管理器中添加的「廣州」按鈕
        gz.setOnClickListener(this);
    }
    @Override
    public void onClick(View view){
        switch(view.getId()){
            case R.id.bj:        //單擊的是「北京」按鈕
                openUrl("101010100");
                break;
            case R.id.sh:        //單擊的是「上海」按鈕
```

```
                openUrl("101020100");
                break;
            case R.id.heb:        //單擊的是「哈爾濱」按鈕
                openUrl("101050101");
                break;
            case R.id.cc:         //單擊的是「長春」按鈕
                openUrl("101060101");
                break;
            case R.id.sy:         //單擊的是「瀋陽」按鈕
                openUrl("101070101");
                break;
            case R.id.gz:         //單擊的是「廣州」按鈕
                openUrl("101280101");
                break;
        }
    }
    //打開網頁的方法
    private void openUrl(String id){
        webView.loadUrl("http://m.weather.com.cn/mweather/" + id +".shtml");   //獲取並顯示天氣預報信息
    }
}
```

‧AndroidManifest.xml 文件代碼為：

```
<?xml version="1.0" encoding="utf-8"?>
<manifest xmlns:android="http://schemas.android.com/apk/res/android"
    package="com.example.webview" >
    <uses-permission android:name="android.permission.INTERNET" />
    <application
        android:allowBackup="true"
        android:icon="@mipmap/ic_launcher"
        android:label="@string/app_name"
        android:supportsRtl="true"
        android:theme="@style/AppTheme" >
        <activity android:name=".MainActivity" >
            <intent-filter>
                <action android:name="android.intent.action.MAIN" />
```

```
                    <category android:name=" android.intent.category.LAUNCHER"
/>
            </intent-filter>
        </activity>
    </application>

</manifest>
```

【練習 6.2】基於 WebService 的手機歸屬地查詢

(1) 運行效果如圖 6.3 所示。

圖 6.3　WebService 練習運行效果圖

(2) 將 ksoap2-android 的 jar 包添加到 Android 工程中,步驟如下:

①在 Google 提供的項目下載網站中下載開發包,網址為:「https://code.google.com/archive/p/ksoap2-android/source」,點擊「Downloads」,選擇「ksoap2-android-assembly-2.4-jar-with-dependencies.jar」,如圖 6.4 所示。

圖 6.4 下載 ksoap2-android

對於下載有難度的讀者，可以直接在本書配套的數字資源包中找到該 jar 包。

②將下載的 ksoap2-android 的 jar 包添加到工程的 lib 目錄下，並右鍵點擊選擇「Add as library」，這樣就將 ksoap2-android 集成到了 Android 項目中了，如圖 6.5 所示。

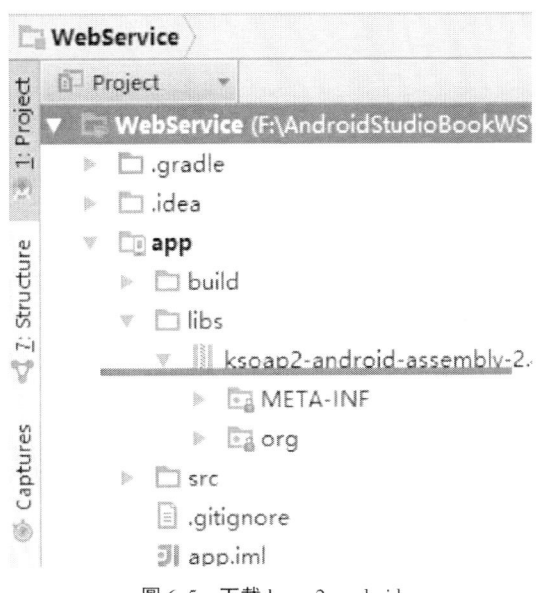

圖 6.5 下載 ksoap2-android

（3）此時就可以根據網上開放的 WebService 服務開發自己的應用程序了，在這一節中將展示如何實現一個手機歸屬地查詢的小應用，我們使用的 WebService 服務地址為：

http://ws.webxml.com.cn/WebServices/MobileCodeWS.asmx

打開網址後點擊「getMobileCodeInfo」進入說明頁，如圖 6.6 所示。

圖 6.6　WebService 服務頁

在 http://ws.webxml.com.cn/WebServices/MobileCodeWS.asmx 後加上「？wsdl」就可以訪問其 wsdl 說明，如圖 6.7 所示。

圖 6.7　wsdl 說明頁

如圖 6.7 所示，我們可以得到幾個很關鍵的點：
①作用域 TargetNameSpace ＝ http://WebXml.com.cn/。
②查詢的方法名為「getMobileCodeInfo」，需要帶上「mobileCode」與「userID」兩個參數。
返回的結果存在「getMobileCodeInfoResult」中。
（4）資源文件佈局如圖 6.8 所示。

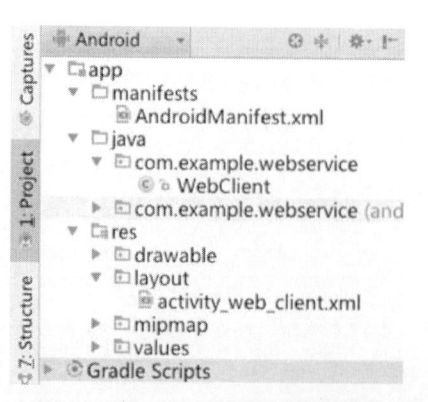

圖 6.8　WebService 練習工程結構圖

· activity_web_client.xml 文件代碼如下：

```xml
<?xml version="1.0" encoding="utf-8"?>
<LinearLayout xmlns:android="http://schemas.android.com/apk/res/android"
    xmlns:tools="http://schemas.android.com/tools"
    android:layout_width="match_parent"
    android:layout_height="match_parent"
    android:paddingBottom="@dimen/activity_vertical_margin"
    android:paddingLeft="@dimen/activity_horizontal_margin"
    android:paddingRight="@dimen/activity_horizontal_margin"
    android:paddingTop="@dimen/activity_vertical_margin"
    android:orientation="vertical"
    tools:context="com.example.webservice.WebClient" >
    <LinearLayout
        android:layout_width="match_parent"
        android:layout_height="wrap_content"
        android:orientation="horizontal" >
        <TextView
            android:layout_width="wrap_content"
            android:layout_height="wrap_content"
            android:text="輸入手機號:" />

        <EditText
            android:layout_width="150dp"
            android:layout_height="wrap_content"
            android:id="@+id/etphone" />

        <Button
            android:layout_width="wrap_content"
            android:layout_height="wrap_content"
            android:text="搜索"
            android:id="@+id/btnsearch" />

    </LinearLayout>

    <TextView
        android:layout_width="wrap_content"
        android:layout_height="wrap_content"
```

```xml
        android:text="查詢結果:"/>

    <TextView
        android:id="@+id/tvinfo"
        android:layout_width="wrap_content"
        android:layout_height="wrap_content"/>

</LinearLayout>
```

・WebClient.java 文件代碼為:

```java
package com.example.webservice;

import android.os.AsyncTask;
import android.support.v7.app.AppCompatActivity;
import android.os.Bundle;
import android.view.View;
import android.widget.Button;
import android.widget.EditText;
import android.widget.TextView;

import org.ksoap2.SoapEnvelope;
import org.ksoap2.SoapFault;
import org.ksoap2.serialization.SoapObject;
import org.ksoap2.serialization.SoapSerializationEnvelope;
import org.ksoap2.transport.HttpTransportSE;
import org.xmlpull.v1.XmlPullParserException;

import java.io.IOException;

public class WebClient extends AppCompatActivity {

    private static final String SERVER_URL = "http://ws.webxml.com.cn/WebServices/MobileCodeWS.asmx?wsdl";
    // 調用的 webservice 命令空間
    private static final String PACE = "http://WebXml.com.cn/";
    // 獲取歸屬地的方法名
    private static final String W_NAME = "getMobileCodeInfo";
```

```java
    private EditText etPhone;
    private Button btnSearch;
    private TextView tvInfo;

    @Override
    protected void onCreate(Bundle savedInstanceState) {
        super.onCreate(savedInstanceState);
        setContentView(R.layout.activity_web_client);
        etPhone = (EditText) findViewById(R.id.etphone);
        btnSearch = (Button) findViewById(R.id.btnsearch);
        tvInfo = (TextView) findViewById(R.id.tvinfo);
        btnSearch.setOnClickListener(new View.OnClickListener() {
            @Override
            public void onClick(View v) {
                String cityName = etPhone.getText().toString();
                if (cityName.length() > 0) {
                    getWeatherInfo(etPhone.getText().toString());
                }
            }
        });
    }

    private void getWeatherInfo(String phoneMum) {
        new AsyncTask<String, Void, String>() {
            @Override
            protected String doInBackground(String... params) {
                String local = "";
                final HttpTransportSE httpSe = new HttpTransportSE(SERVER_URL);
                httpSe.debug = true;
                SoapObject soapObject = new SoapObject(PACE, W_NAME);
                soapObject.addProperty("mobileCode", params[0]);
                soapObject.addProperty("userID", "");
                final SoapSerializationEnvelope serializa = new SoapSerializationEnvelope(
                        SoapEnvelope.VER10);
                serializa.setOutputSoapObject(soapObject);
```

```
                    serializa.dotNet = true;
                    // 獲取返回信息
                    try {
                        httpSe.call(PACE + W_NAME, serializa);
                        if (serializa.getResponse() ! = null) {
                            SoapObject result = (SoapObject) serializa.bodyIn;
                            local = result.getProperty ( "
getMobileCodeInfoResult" ).toString();
                        }
                    }
                    catch (XmlPullParserException e) {
                        e.printStackTrace();
                    } catch (SoapFault soapFault) {
                        soapFault.printStackTrace();
                    } catch (IOException e) {
                        e.printStackTrace();
                    }
                    return local;
                }

                @Override
                protected void onPostExecute(String result) {
                    tvInfo.setText(result);
                }
            }.execute(phoneMum);
        }
    }
```

・AndroidManifest.xml 文件代碼為:

```
<?xml version="1.0" encoding="utf-8"?>
<manifest xmlns:android="http://schemas.android.com/apk/res/android"
    package="com.example.webservice">

    <uses-permission android:name="android.permission.INTERNET"/>

    <application
        android:allowBackup="true"
```

```
            android:icon="@mipmap/ic_launcher"
            android:label="@string/app_name"
            android:supportsRtl="true"
            android:theme="@style/AppTheme">
            <activity android:name="com.example.webservice.WebClient">
                <intent-filter>
                    <action android:name="android.intent.action.MAIN" />

                    <category android:name="android.intent.category.LAUNCHER" />
                </intent-filter>
            </activity>
        </application>

</manifest>
```

6.4 擴展練習

(1)編寫 Android 程序，實現使用系統內置瀏覽器打開指定網頁。
(2)編寫 Android 程序，實現從指定網站下載文件。

6.5 實驗報告

(1)每人一份實驗報告，統一用學校提供的 A4 幅面的實驗報告冊書寫或用 A4 的紙打印。如果打印必須有以下格式的表頭：

實驗課程					
實驗名稱					
實驗時間	學年	學期	週	星期	第　節
學生姓名	學號		班級		
同組姓名	學號		班級		
實驗地點	設備號		指導教師		

(2)實驗內容的主要結果及對結果的分析。
(3)實驗過程中你所遇到的問題的解決辦法。
(4)心得體會、意見和建議。

6.6 實驗成績考核

(1) 考勤占 10%。
(2) 相互協作完成實驗任務占 40%。
(3) 實驗報告占 50%。

實驗七　SQLite 和 SQLiteDatabase 的使用

7.1　實驗目的

本次實驗的目的是讓大家熟悉 Android 中對數據庫進行相關的數據操作。SQLite-Database 是在 Android 中數據庫操作使用最頻繁的一個類。通過它可以實現數據庫的創建或打開、創建表、插入數據、刪除數據、查詢數據、修改數據等操作。

7.2　實驗要求

（1）實現便簽管理小例程。
（2）創建項目並熟悉文件目錄結構。
（3）掌握實現便簽增刪改查功能的實驗步驟。

7.3　實驗內容

【練習 7.1】便簽管理小例程

（1）程序運行效果如圖 7.1、圖 7.2、圖 7.3、圖 7.4 所示。

圖 7.1　Activity_main.xml（啓動窗體）運行效果圖

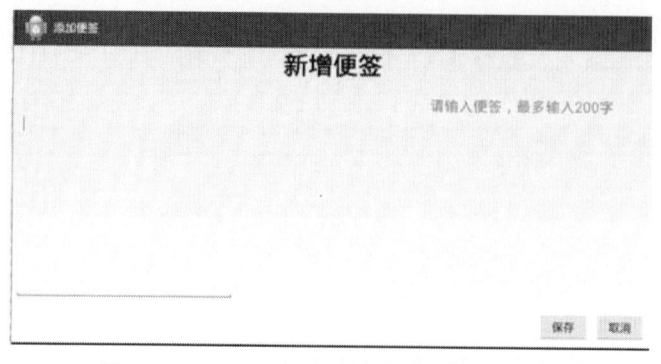

圖 7.2　Insertinfo.xml(新增便簽窗體)運行效果圖

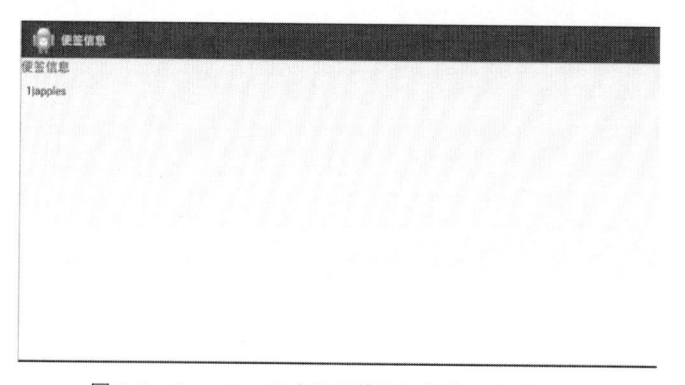

圖 7.3　showinfo.xml(查看便簽信息窗體)運行效果圖

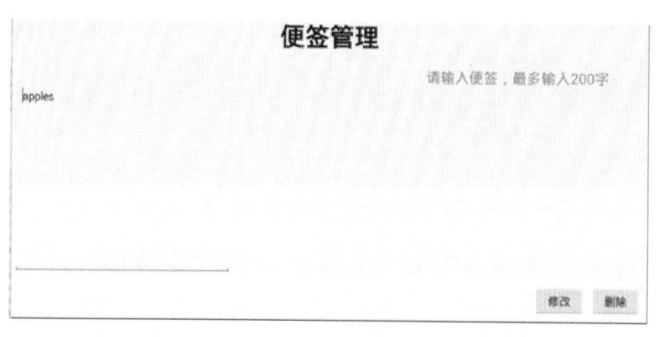

圖 7.4　manageflag.xml(便簽管理窗體)運行效果圖

(2)資源文件佈局如圖 7.5 所示。

```
Android                    ⊙ ÷ ※ ⊦
▼ app
    ▼ manifests
        AndroidManifest.xml
    ▼ java
        ▼ com.wenlong.DBLab
            ▼ activity
                InsertFlag
                MainActivity
                ManageFlag
                ShowInfo
            ▼ DAO
                DBOpenHelper
                FlagDao
            ▼ model
                flag
    ▼ res
        ▶ drawable
        ▼ layout
            activity_main.xml
            insertinfo.xml
            manageflag.xml
            showinfo.xml
        ▶ menu
        ▶ values
    ▶ Gradle Scripts
```

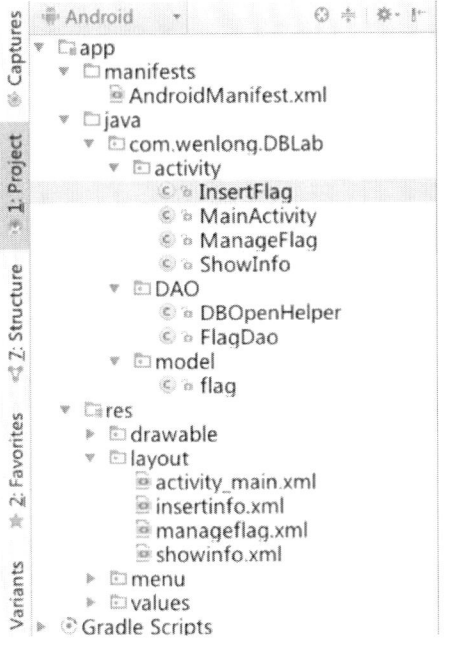

圖 7.5　資源文件佈局圖

- activity_main.xml(啓動窗體)文件代碼如下：

```
<LinearLayout xmlns:android="http://schemas.android.com/apk/res/android"
    android:layout_width="fill_parent"
    android:layout_height="fill_parent"
    android:orientation="vertical" >
<LinearLayout android:id="@+id/linearLayout1"
        android:layout_height="wrap_content"
        android:layout_width="match_parent"
        android:orientation="vertical"
        android:layout_weight="0.06" >
    <RelativeLayout android:layout_height="wrap_content"
        android:layout_width="match_parent" >
    <Button android:text="便簽信息"
        android:id="@+id/btnflaginfo"
        android:layout_width="wrap_content"
        android:layout_height="wrap_content"
        android:textSize="20dp"
        android:textColor="#8C6931"
```

```xml
        />
        <Button android:text="添加便簽"
            android:id="@+id/btninsertinfo"
            android:layout_width="wrap_content"
            android:layout_height="wrap_content"
            android:layout_toRightOf="@id/btnflaginfo"
            android:textSize="20dp"
            android:textColor="#8C6931"
            />
    </RelativeLayout>
    </LinearLayout>
</LinearLayout>
```

・Insertinfo.xml(新增便簽窗體)文件代碼如下:

```xml
<?xml version="1.0" encoding="utf-8"?>
<LinearLayout xmlns:android="http://schemas.android.com/apk/res/android"
    android:id="@+id/itemflag"
    android:orientation="vertical"
    android:layout_width="fill_parent"
    android:layout_height="fill_parent"
    >
    <LinearLayout
        android:orientation="vertical"
        android:layout_width="fill_parent"
        android:layout_height="fill_parent"
        android:layout_weight="3"
        >
        <TextView
            android:layout_width="wrap_content"
            android:layout_gravity="center"
            android:gravity="center_horizontal"
            android:text="新增便簽"
            android:textSize="40sp"
            android:textColor="#000000"
            android:textStyle="bold"
            android:layout_height="wrap_content" />
    </LinearLayout>
```

```xml
<LinearLayout
    android:orientation="vertical"
    android:layout_width="fill_parent"
    android:layout_height="fill_parent"
    android:layout_weight="1"
    >
    <RelativeLayout android:layout_width="fill_parent"
        android:layout_height="fill_parent"
        android:padding="5dp"
        >
        <TextView android:layout_width="350dp"
        android:id="@+id/tvFlag"
        android:textSize="23sp"
        android:text="請輸入便簽,最多輸入200字"
        android:textColor="#8C6931"
        android:layout_alignParentRight="true"
        android:layout_height="wrap_content"
        />
        <EditText
        android:id="@+id/txtFlag"
        android:layout_width="350dp"
        android:layout_height="400dp"
        android:layout_below="@id/tvFlag"
        android:gravity="top"
        android:singleLine="false"
        />
    </RelativeLayout>
</LinearLayout>
<LinearLayout
    android:orientation="vertical"
    android:layout_width="fill_parent"
    android:layout_height="fill_parent"
    android:layout_weight="3"
    >
    <RelativeLayout android:layout_width="fill_parent"
        android:layout_height="fill_parent"
        android:padding="10dp"
        >
```

```xml
<Button
    android:id="@+id/btnflagCancel"
    android:layout_width="80dp"
    android:layout_height="wrap_content"
    android:layout_alignParentRight="true"
    android:layout_marginLeft="10dp"
    android:text="取消"
    />
<Button
    android:id="@+id/btnflagSave"
    android:layout_width="80dp"
    android:layout_height="wrap_content"
    android:layout_toLeftOf="@id/btnflagCancel"
    android:text="保存"
    android:maxLength="200"
    />
        </RelativeLayout>
    </LinearLayout>
</LinearLayout>
```

・showinfo.xml(查看便簽信息窗體)文件代碼如下：

```xml
<?xml version="1.0" encoding="utf-8"?>
<LinearLayout xmlns:android="http://schemas.android.com/apk/res/android"
    android:layout_width="match_parent"
    android:layout_height="match_parent"
    android:orientation="vertical" >

    <TextView
        android:id="@+id/textView1"
        android:layout_width="wrap_content"
        android:layout_height="wrap_content"
        android:textSize="20dp"
        android:text="便簽信息" />
    <LinearLayout android:id="@+id/linearLayout2"
        android:layout_height="wrap_content"
        android:layout_width="match_parent"
        android:orientation="vertical"
```

```xml
        android:layout_weight="0.94">
        <ListView android:id="@+id/lvinfo"
            android:layout_width="match_parent"
            android:layout_height="match_parent"
            android:scrollbarAlwaysDrawVerticalTrack="true"
            />
    </LinearLayout>
</LinearLayout>
```

· manageflag.xml(便簽管理窗體)文件代碼如下：

```xml
<?xml version="1.0" encoding="utf-8"?>
<LinearLayout xmlns:android="http://schemas.android.com/apk/res/android"
    android:id="@+id/flagmanage"
    android:orientation="vertical"
    android:layout_width="fill_parent"
    android:layout_height="fill_parent"
    >
    <LinearLayout
        android:orientation="vertical"
        android:layout_width="fill_parent"
        android:layout_height="fill_parent"
        android:layout_weight="3"
        >
        <TextView
            android:layout_width="wrap_content"
            android:layout_gravity="center"
            android:gravity="center_horizontal"
            android:text="便簽管理"
            android:textSize="40sp"
            android:textColor="#000000"
            android:textStyle="bold"
            android:layout_height="wrap_content"/>
    </LinearLayout>
    <LinearLayout
        android:orientation="vertical"
        android:layout_width="fill_parent"
        android:layout_height="fill_parent"
```

```xml
        android:layout_weight="1"
        >
        <RelativeLayout android:layout_width="fill_parent"
            android:layout_height="fill_parent"
            android:padding="5dp"
            >
            <TextView android:layout_width="350dp"
                android:id="@+id/tvFlagManage"
                android:textSize="23sp"
                android:text="請輸入便簽,最多輸入200字"
                android:textColor="#8C6931"
                android:layout_alignParentRight="true"
                android:layout_height="wrap_content"
                />
            <EditText
                android:id="@+id/txtFlagManage"
                android:layout_width="350dp"
                android:layout_height="400dp"
                android:layout_below="@id/tvFlagManage"
                android:gravity="top"
                android:singleLine="false"
                />
        </RelativeLayout>
</LinearLayout>
<LinearLayout
    android:orientation="vertical"
    android:layout_width="fill_parent"
    android:layout_height="fill_parent"
    android:layout_weight="3"
    >
    <RelativeLayout android:layout_width="fill_parent"
        android:layout_height="fill_parent"
        android:padding="10dp"
        >
        <Button
            android:id="@+id/btnFlagManageDelete"
            android:layout_width="80dp"
            android:layout_height="wrap_content"
```

```xml
        android:layout_alignParentRight="true"
        android:layout_marginLeft="10dp"
        android:text="删除"
        />
    <Button
        android:id="@+id/btnFlagManageEdit"
        android:layout_width="80dp"
        android:layout_height="wrap_content"
        android:layout_toLeftOf="@id/btnFlagManageDelete"
        android:text="修改"
        android:maxLength="200"
        />
    </RelativeLayout>
</LinearLayout>
</LinearLayout>
```

- MainActivity.java 文件代码如下:

```java
package com.wenlong.DBLab.activity;

import android.os.Bundle;
import android.app.Activity;
import android.content.Intent;
import android.view.Menu;
import android.view.View;
import android.view.View.OnClickListener;
import android.widget.Button;

public class MainActivity extends Activity {
Button btnflaginfo,btninsertinfo;
    @Override
    protected void onCreate(Bundle savedInstanceState) {
        super.onCreate(savedInstanceState);
        setContentView(R.layout.activity_main);
        btnflaginfo=(Button)findViewById(R.id.btnflaginfo);
        btninsertinfo=(Button)findViewById(R.id.btninsertinfo);

        btnflaginfo.setOnClickListener(new OnClickListener() {
```

```
            @Override
            public void onClick(View v){
                Intent intent = new Intent(MainActivity.this, ShowInfo.class);
                startActivity(intent);
            }

        });
        btninsertinfo.setOnClickListener(new OnClickListener(){

            @Override
            public void onClick(View v){
                Intent intent = new Intent(MainActivity.this, InsertFlag.class);
                startActivity(intent);
            }

        });
    }

    @Override
    public boolean onCreateOptionsMenu(Menu menu){
        // Inflate the menu; this adds items to the action bar if it is present.
        getMenuInflater().inflate(R.menu.main, menu);
        return true;
    }

}
```

· InsertFlag.java 文件代碼如下：

```
package com.wenlong.DBLab.activity;

import android.app.Activity;
import android.os.Bundle;
import android.view.View;
import android.view.View.OnClickListener;
import android.widget.Button;
import android.widget.EditText;
```

```java
import android.widget.Toast;

import com.wenlong.DBLab.DAO.FlagDao;
import com.wenlong.DBLab.model.flag;

public class InsertFlag extends Activity {
    EditText txtFlag;// 創建 EditText 組件對象
    Button btnflagSaveButton;// 創建 Button 組件對象
    Button btnflagCancelButton;// 創建 Button 組件對象

    @Override
    protected void onCreate(Bundle savedInstanceState) {
        super.onCreate(savedInstanceState);
        setContentView(R.layout.insertinfo);
        txtFlag = (EditText)findViewById(R.id.txtFlag);
        btnflagSaveButton = (Button)findViewById(R.id.btnflagSave);
        btnflagCancelButton = (Button)findViewById(R.id.btnflagCancel);
        btnflagSaveButton.setOnClickListener(new OnClickListener() {

            @Override
            public void onClick(View v) {
                String strFlag = txtFlag.getText().toString();// 獲取便簽文本框的值
                if (! strFlag.isEmpty()) {// 判斷獲取
                    FlagDao flagDAO = new FlagDao(InsertFlag.this);// 創建 FlagDAO 對象
                    flag flag = new flag(
                        flagDAO.getMaxId() + 1, strFlag);// 創建 Tb_flag 對象
                    flagDAO.add(flag);// 添加便簽信息
                    // 彈出信息提示
                    Toast.makeText(InsertFlag.this, "‖新增便簽‖數據添加成功！",
                        Toast.LENGTH_SHORT).show();
                } else {
                    Toast.makeText(InsertFlag.this, "請輸入便簽！",
                        Toast.LENGTH_SHORT).show();
                }
            }
        });
    }
}
```

```java
        });
        btnflagCancelButton.setOnClickListener( new OnClickListener( ) {
            @Override
            public void onClick( View v ) {
                finish( );
            }
        });
    }
}
```

·ShowInfo.java 文件代碼如下：

```java
package com.wenlong.DBLab.activity;

import android.app.Activity;
import android.content.Intent;
import android.os.Bundle;
import android.view.View;
import android.widget.AdapterView;
import android.widget.AdapterView.OnItemClickListener;
import android.widget.ArrayAdapter;
import android.widget.ListView;
import android.widget.TextView;

import com.wenlong.DBLab.DAO.FlagDao;
import com.wenlong.DBLab.model.flag;

import java.util.List;

public class ShowInfo extends Activity {
    public static final String FLAG = "id";// 定義一個常量,用來作為請求碼
    ListView lvinfo;// 創建 ListView 對象
    String[] strInfos = null;// 定義字符串數組,用來存儲收入信息
    ArrayAdapter<String> arrayAdapter = null;// 創建 ArrayAdapter 對象
    @Override
    protected void onCreate( Bundle savedInstanceState ) {
        super.onCreate( savedInstanceState );
        setContentView( R.layout.showinfo );
```

```java
lvinfo = (ListView)findViewById(R.id.lvinfo);
FlagDao flaginfo = new FlagDao(ShowInfo.this);// 創建 FlagDAO 對象
// 獲取所有便簽信息,並存儲到 List 泛型集合中
List<flag> listFlags = flaginfo.getScrollData(0,
        (int)flaginfo.getCount());
strInfos = new String[listFlags.size()];// 設置字符串數組的長度
int n = 0;// 定義一個開始標示
for(flag tb_flag : listFlags)
{
    // 將便簽相關信息組合成一個字符串,存儲到字符串數組的相應位置
    strInfos[n] = tb_flag.getid() + "|" + tb_flag.getFlag();
    if(strInfos[n].length() > 15)// 判斷便簽信息的長度是否大於 15
        strInfos[n] = strInfos[n].substring(0, 15) + "……";// 將位置大於 15 之後的字符串用……代替
    n++;// 標示加 1
}
arrayAdapter = new ArrayAdapter<String>(this,
        android.R.layout.simple_list_item_1, strInfos);
lvinfo.setAdapter(arrayAdapter);
lvinfo.setOnItemClickListener(new OnItemClickListener()
{
    @Override
    public void onItemClick(AdapterView<?> parent, View view, int position,
                            long id) {
        String strInfo = String.valueOf(((TextView)view).getText());// 記錄單擊的項信息
        String strid = strInfo.substring(0, strInfo.indexOf('|'));// 從項信息中截取編號
        Intent intent = null;// 創建 Intent 對象
        intent = new Intent(ShowInfo.this, ManageFlag.class);// 使用 FlagManage 窗口初始化 Intent 對象
        intent.putExtra(FLAG, strid);// 設置要傳遞的數據
        startActivity(intent);// 執行 Intent,打開相應的 Activity
    }
});
}
}
```

- ManageFlag.java 文件代碼如下：

```java
package com.wenlong.DBLab.activity;

import android.app.Activity;
import android.content.Intent;
import android.os.Bundle;
import android.view.View;
import android.view.View.OnClickListener;
import android.widget.Button;
import android.widget.EditText;
import android.widget.Toast;

import com.wenlong.DBLab.DAO.FlagDao;
import com.wenlong.DBLab.model.flag;
public class ManageFlag extends Activity {
    EditText txtFlag;// 創建 EditText 對象
    Button btnEdit, btnDel;// 創建兩個 Button 對象
    String strid;// 創建字符串,表示便簽的 id
    @Override
    protected void onCreate(Bundle savedInstanceState) {
        super.onCreate(savedInstanceState);
        setContentView(R.layout.manageflag);
        txtFlag = (EditText)findViewById(R.id.txtFlagManage);
        btnEdit = (Button)findViewById(R.id.btnFlagManageEdit);
        btnDel = (Button)findViewById(R.id.btnFlagManageDelete);
        Intent intent = getIntent();// 創建 Intent 對象
        Bundle bundle = intent.getExtras();// 獲取便簽 id
        strid = bundle.getString(ShowInfo.FLAG);// 將便簽 id 轉換為字符串
        final FlagDao flagDAO = new FlagDao(ManageFlag.this);// 創建 FlagDAO 對象
        txtFlag.setText(flagDAO.find(Integer.parseInt(strid)).getFlag());
        // 為修改按鈕設置監聽事件
        btnEdit.setOnClickListener(new OnClickListener() {
            @Override
            public void onClick(View v) {
                flag tb_flag = new flag();// 創建 Tb_flag 對象
                tb_flag.setid(Integer.parseInt(strid));// 設置便簽 id
```

```
            tb_flag.setFlag(txtFlag.getText().toString());// 設置便簽值
            flagDAO.update(tb_flag);// 修改便簽信息
            // 彈出信息提示
            Toast.makeText(ManageFlag.this, "[便簽數據]修改成功!",
                    Toast.LENGTH_SHORT).show();

        }
    });
    // 為刪除按鈕設置監聽事件
    btnDel.setOnClickListener(new OnClickListener() {

        @Override
        public void onClick(View v) {
            flagDAO.detele(Integer.parseInt(strid));// 根據指定的 id 刪除便簽信息
            Toast.makeText(ManageFlag.this, "[便簽數據]刪除成功!",
                    Toast.LENGTH_SHORT).show();

        }
    });
  }
}
```

· DBOpenHelper.java 文件代碼如下:

```
package com.wenlong.DBLab.DAO;

import android.content.Context;
import android.database.sqlite.SQLiteDatabase;
import android.database.sqlite.SQLiteOpenHelper;

public class DBOpenHelper extends SQLiteOpenHelper {

    private static final int VERSION = 1;// 定義數據庫版本號
    private static final String DBNAME = "flag.db";// 定義數據庫名
    public DBOpenHelper(Context context) {
        super(context, DBNAME, null, VERSION);
    }
    @Override
```

```java
public void onCreate(SQLiteDatabase db) // 創建數據庫
{
    db.execSQL("create table tb_flag (_id integer primary key, flag varchar(200))");// 創建便簽信息表
}
@Override
public void onUpgrade(SQLiteDatabase db, int oldVersion, int newVersion) // 覆寫基類的 onUpgrade 方法,以便數據庫版本更新
{
}
}
```

·FlagDao.java 文件代碼如下:

```java
package com.wenlong.DBLab.DAO;
import java.util.ArrayList;
import java.util.List;
import android.content.Context;
import android.database.Cursor;
import android.database.sqlite.SQLiteDatabase;
import com.wenlong.DBLab.DAO.DBOpenHelper;
import com.wenlong.DBLab.model.flag;
public class FlagDao {
    private DBOpenHelper helper;// 創建 DBOpenHelper 對象
    private SQLiteDatabase db;// 創建 SQLiteDatabase 對象

    public FlagDao(Context context)// 定義構造函數
    {
        helper = new DBOpenHelper(context);// 初始化 DBOpenHelper 對象
    }
    /**
     * 添加便簽信息
     *
     * @param tb_flag
     */
    public void add(flag flag) {
        db = helper.getWritableDatabase();// 初始化 SQLiteDatabase 對象
        db.execSQL("insert into tb_flag (_id,flag) values (?,?)", new Object[] {
```

            flag.getid( ) , flag.getFlag( ) }) ;// 執行添加便簽信息操作
}
/ * *
    * 更新便簽信息
    *
    * @ param tb_flag
    */
public void update(flag tb_flag) {
    db = helper.getWritableDatabase( ) ;// 初始化 SQLiteDatabase 對象
    db.execSQL("update tb_flag set flag = ? where _id = ?", new Object[ ] {
        tb_flag.getFlag( ) , tb_flag.getid( ) }) ;// 執行修改便簽信息操作
}
/ * *
    * 查找便簽信息
    *
    * @ param id
    * @ return
    */
public flag find(int id) {
    db = helper.getWritableDatabase( ) ;// 初始化 SQLiteDatabase 對象
    Cursor cursor = db.rawQuery(
            "select _id,flag from tb_flag where _id = ?",
            new String[ ] { String.valueOf(id) }) ;// 根據編號查找便簽信息,
並存儲到 Cursor 類中
    if (cursor.moveToNext( ))// 遍歷查找到的便簽信息
    {
        // 將遍歷到的便簽信息存儲到 Tb_flag 類中
        return new flag(cursor.getInt(cursor.getColumnIndex("_id")),
            cursor.getString(cursor.getColumnIndex("flag")));
    }
    return null;// 如果沒有信息,則返回 null
}
/ * *
    * 刪除便簽信息
    *
    * @ param ids
    */

```java
public void detele(Integer... ids) {
    if (ids.length > 0)// 判斷是否存在要刪除的 id
    {
        StringBuffer sb = new StringBuffer();// 創建 StringBuffer 對象
        for (int i = 0; i < ids.length; i++)// 遍歷要刪除的 id 集合
        {
            sb.append('?').append(',');// 將刪除條件添加到 StringBuffer 對象中
        }
        sb.deleteCharAt(sb.length() - 1);// 去掉最後一個「,」字符
        db = helper.getWritableDatabase();// 創建 SQLiteDatabase 對象
        // 執行刪除便簽信息操作
        db.execSQL("delete from tb_flag where _id in (" + sb + ")",
            (Object[]) ids);
    }
}

/**
 * 獲取便簽信息
 *
 * @param start
 *            起始位置
 * @param count
 *            每頁顯示數量
 * @return
 */
public List<flag> getScrollData(int start, int count) {
    List<flag> lisTb_flags = new ArrayList<flag>();// 創建集合對象
    db = helper.getWritableDatabase();// 初始化 SQLiteDatabase 對象
    // 獲取所有便簽信息
    Cursor cursor = db.rawQuery("select * from tb_flag limit ?,?",
        new String[] { String.valueOf(start), String.valueOf(count) });
    while (cursor.moveToNext())// 遍歷所有的便簽信息
    {
        // 將遍歷到的便簽信息添加到集合中
        lisTb_flags.add(new flag(cursor.getInt(cursor
            .getColumnIndex("_id")), cursor.getString(cursor
            .getColumnIndex("flag"))));
```

            }
            return lisTb_flags;// 返回集合
    }
    /**
     * 獲取總記錄數
     *
     * @return
     */
    public long getCount() {
        db = helper.getWritableDatabase();// 初始化 SQLiteDatabase 對象
        Cursor cursor = db.rawQuery("select count(_id) from tb_flag", null);// 獲取便簽信息的記錄數
        if (cursor.moveToNext())// 判斷 Cursor 中是否有數據
        {
            return cursor.getLong(0);// 返回總記錄數
        }
        return 0;// 如果沒有數據,則返回 0
    }
    /**
     * 獲取便簽最大編號
     *
     * @return
     */
    public int getMaxId() {
        db = helper.getWritableDatabase();// 初始化 SQLiteDatabase 對象
        Cursor cursor = db.rawQuery("select max(_id) from tb_flag", null);// 獲取便簽信息表中的最大編號
        while (cursor.moveToLast()) {// 訪問 Cursor 中的最後一條數據
            return cursor.getInt(0);// 獲取訪問到的數據,即最大編號
        }
        return 0;// 如果沒有數據,則返回 0
    }
}

・flag.java 文件代碼如下：

```java
package com.wenlong.DBLab.model;

public class flag {
    private int _id;// 存儲便簽編號
    private String flag;// 存儲便簽信息

    public flag()// 默認構造函數
    {
        super();
    }
    // 定義有參構造函數,用來初始化便簽信息實體類中的各個字段
    public flag(int id, String flag) {
        super();
        this._id = id;// 為便簽號賦值
        this.flag = flag;// 為便簽信息賦值
    }
    public int getid()// 設置便簽編號的可讀屬性
    {
        return _id;
    }
    public void setid(int id)// 設置便簽編號的可寫屬性
    {
        this._id = id;
    }
    public String getFlag()// 設置便簽信息的可讀屬性
    {
        return flag;
    }
    public void setFlag(String flag)// 設置便簽信息的可寫屬性
    {
        this.flag = flag;
    }
}
```

· AndroidManifest.xml 配置文件代碼如下：

```xml
<?xml version="1.0" encoding="utf-8"?>
<manifest xmlns:android="http://schemas.android.com/apk/res/android"
    package="com.wenlong.DBLab.activity"
    android:versionCode="1"
    android:versionName="1.0" >

    <uses-sdk
        android:minSdkVersion="8"
        android:targetSdkVersion="18" />

    <application
        android:allowBackup="true"
        android:icon="@drawable/ic_launcher"
        android:label="@string/app_name"
        android:theme="@style/AppTheme" >
        <activity
            android:name="com.wenlong.DBLab.activity.MainActivity"
            android:label="@string/app_name" >
            <intent-filter>
                <action android:name="android.intent.action.MAIN" />

                <category android:name="android.intent.category.LAUNCHER" />
            </intent-filter>
        </activity>
        <activity
            android:label="便簽信息"
            android:icon="@drawable/ic_launcher"
            android:name="com.wenlong.DBLab.activity.ShowInfo" >
        </activity>
        <activity
            android:label="添加便簽"
            android:icon="@drawable/ic_launcher"
            android:name="com.wenlong.DBLab.activity.InsertFlag" >
        </activity>
        <activity
```

```
            android:label = "便簽管理"
            android:icon = "@drawable/ic_launcher"
            android:name = "com.wenlong.DBLab.activity.ManageFlag" >
        </activity>
    </application>

</manifest>
```

7.4 擴展練習

編寫 Android 項目,實現商品庫存數據庫管理小系統。

7.5 實驗報告

(1)每人一份實驗報告,統一用學校提供的 A4 幅面的實驗報告冊書寫或用 A4 的紙打印。如果打印必須有以下格式的表頭:

實驗課程						
實驗名稱						
實驗時間		學年	學期	周	星期	第　節
學生姓名		學號		班級		
同組姓名		學號		班級		
實驗地點		設備號		指導教師		

(2)實驗內容的主要結果及對結果的分析。
(3)實驗過程中你所遇到的問題的解決辦法。
(4)心得體會、意見和建議。

7.6 實驗成績考核

(1)考勤占 10%。
(2)相互協作完成實驗任務占 40%。
(3)實驗報告占 50%。

實驗八　使用 GPS 與百度地圖

8.1　實驗目的

我們能夠通過 GPS 服務獲取精確的定位信息,再同時結合地圖,為用戶帶來更多的定位服務,例如地圖定位、線路規劃、導航等功能。現在的互聯網公司在開發其地圖應用的同時,也為開發者開放了 SDK,例如谷歌地圖、百度地圖與高德地圖等。我們在編程時僅需導入庫文件,然後調用接口,就可以快捷地將這些地圖服務功能集成到自己的應用中。

8.2　實驗要求

使用百度地圖所提供的 SDK,並實現一個軌跡記錄的小應用。

8.3　實驗內容

1. 獲取百度地圖 SDK

百度地圖為開發者提供了完善的服務指南,進入百度開放平臺:http://lbsyun.baidu.com/,選擇「開發」→「Android 開發」→「Android 定位 SDK」,進入頁面後,就可以開始我們的準備工作了。

(1) 獲取秘鑰。

在使用百度地圖 SDK 提供的各種功能之前,需要獲取百度地圖移動版的開發密鑰,並且在工程中進行設置。下面介紹設置流程:

登錄 API 控製臺,點擊「創建應用」,如圖 8.1 所示。

图 8.1　創建應用

在進入頁面後選擇應用類型,然後完善各類信息,如圖 8.2 所示。

圖 8.2　填寫信息

在這個頁面中需要填寫數字簽名(SHA1),以前在 Eclipse 中可以很方便地查看到數字簽名,而在 Android Studio 中卻不能直接查看。我們可以在 Android Studio 中打開 Terminal,然後定位到秘鑰所在的文件目錄下(通常默認在 C:\Users\Xxx\.android 中),輸入:「keytool -v -list -keystore debug.keystore」,其中的「debug.keystore」就是您所使用的調試秘鑰,然後就可以看到數字簽名,如圖 8.3 所示。

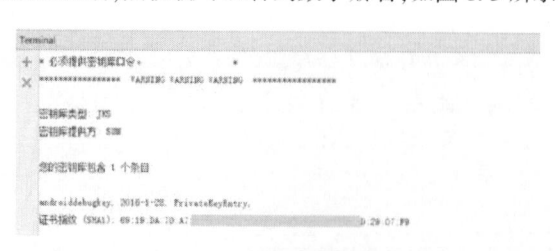

圖 8.3　生成數字簽名

如果打包生成 APK,則需要點擊「Build」→「Generate Signed」,生成秘鑰之後再用此方法獲得數字簽名。

表格中的「包名」可以在 AndroidManifest.xml 文件中查看,如圖 8.4 所示。

圖 8.4　查看包名

將對應的 SHA1 碼與包名填入即可創建應用,獲得 appkey,如圖 8.5 所示。

圖 8.5　生成 appkey

(2)工程配置。

在下載好 SDK 庫文件之後,就可以開始工程配置了。步驟如下:

首先下載解壓開發包,將工程視圖改為 ProjectFile,將開發包中的 libs 文件複製到項目名/app/libs 文件夾中,Android Studio 會自動加載 jar 包與 so 文件。然後右鍵選擇 Add As Library,導入工程中。然後在 src/main 文件目錄下創建 jniLibs 文件夾,放入對應平臺的 so 文件,工程目錄如圖 8.6 所示。

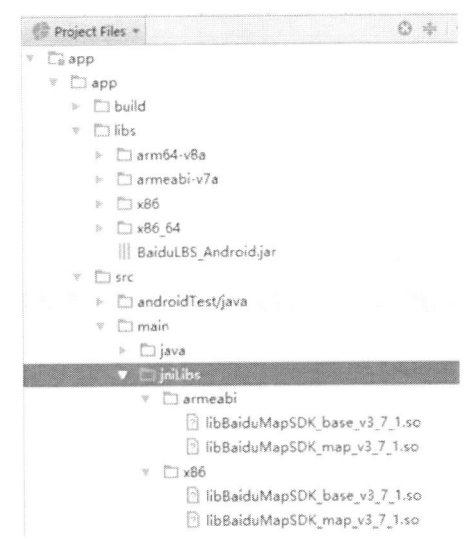

圖 8.6　工程目錄

注意如果需要在 Genymotion 模擬器上調試該程序,還需要將 x86 平臺中的 so 文件導入 jniLibs 中,如圖 8.6 所示。

最後在打包混淆的時候,需要注意與地圖 SDK 相關的方法不可被混淆。混淆方法如下:

-keep classcom.baidu.＊＊｛＊;｝

-keep classvi.com.＊＊｛＊;｝

-dontwarncom.baidu.＊＊

2. 使用百度地圖定位

下文將介紹如何加載百度地圖，並調用百度定位SDK進行精確定位，與之前準備工作稍有不同的是，這次還需要將百度定位的SDK加入工程目錄中，才能使用百度地圖定位服務。

（1）首先配置Manifest文件，添加appkey與所需權限。

```xml
<?xml version="1.0" encoding="utf-8"?>
<manifest xmlns:android="http://schemas.android.com/apk/res/android"
    package="com.example.bdmap">
<!--添加需要的權限-->
<uses-permission android:name="android.permission.ACCESS_NETWORK_STATE"/>
<uses-permission android:name="android.permission.INTERNET"/>
<uses-permission android:name="com.android.launcher.permission.READ_SETTINGS"/>
<uses-permission android:name="android.permission.WAKE_LOCK"/>
<uses-permission android:name="android.permission.CHANGE_WIFI_STATE"/>
<uses-permission android:name="android.permission.ACCESS_WIFI_STATE"/>
<uses-permission android:name="android.permission.GET_TASKS"/>
<uses-permission android:name="android.permission.WRITE_EXTERNAL_STORAGE"/>
<uses-permission android:name="android.permission.WRITE_SETTINGS"/>
<!-- 這個權限用於進行網路定位 -->
<uses-permission android:name="android.permission.ACCESS_COARSE_LOCATION"/>
<!-- 這個權限用於訪問GPS定位 -->
<uses-permission android:name="android.permission.ACCESS_FINE_LOCATION"/>
<!-- 用於訪問wifi網路信息，wifi信息會用於進行網路定位 -->
<uses-permission android:name="android.permission.ACCESS_WIFI_STATE"/>
    <application
        android:allowBackup="true"
        android:icon="@mipmap/ic_launcher"
        android:label="@string/app_name"
```

```xml
        android:supportsRtl="true"
        android:theme="@style/AppTheme">
        <!--添加appkey-->
            <meta-data
                android:name="com.baidu.lbsapi.API_KEY"
                android:value="7znnXUp2G3LIMQ8g3z7OSKZp" />
            <activity android:name=".BaiduActivity">
                <intent-filter>
                    <action android:name="android.intent.action.MAIN" />
                    <category android:name="android.intent.category.LAUNCHER" />
                </intent-filter>
            </activity>
        <!--添加百度服務-->
            <service
                android:name="com.baidu.location.f"
                android:enabled="true"
                android:process=":remote">
            </service>
        </application>
</manifest>
```

(2)然後定義主Activity。

```java
package com.example.bdmap;
    import android.support.v7.app.AppCompatActivity;
    import android.os.Bundle;
    import com.baidu.location.BDLocation;
    import com.baidu.location.BDLocationListener;
    import com.baidu.location.LocationClient;
    import com.baidu.location.LocationClientOption;
    import com.baidu.mapapi.SDKInitializer;
    import com.baidu.mapapi.map.BaiduMap;
    import com.baidu.mapapi.map.MapStatus;
    import com.baidu.mapapi.map.MapStatusUpdateFactory;
    import com.baidu.mapapi.map.MapView;
    import com.baidu.mapapi.map.MyLocationData;
    import com.baidu.mapapi.model.LatLng;
```

```java
public class BaiduActivity extends AppCompatActivity {
    MapView mMapView = null;
    BaiduMap mBaiduMap;
    LocationClient mLocClient;
    boolean isFirstLoc = true; //是否首次定位
    public MyLocationListenner myListener = new MyLocationListenner( );
    @Override
    protected void onCreate( Bundle savedInstanceState) {
    super.onCreate( savedInstanceState);
        SDKInitializer.initialize( getApplicationContext( ) );
        setContentView( R.layout.activity_baidu);
        mMapView = ( MapView) findViewById( R.id.bmapView);
        mBaiduMap = mMapView.getMap( );
        //開啟定位圖層
        mBaiduMap.setMyLocationEnabled( true);
        //定位初始化
        mLocClient = new LocationClient( this);
        mLocClient.registerLocationListener( myListener);
        LocationClientOption option = newLocationClientOption( );
        option.setOpenGps( true); //打開 gps
        option.setCoorType( "bd09ll" ); //設置坐標類型
option.setScanSpan( 1000);
        mLocClient.setLocOption( option);
        mLocClient.start( );
    }
//定位 SDK 監聽函數
public class MyLocationListenner implements BDLocationListener {
    
    @Override
    public voidonReceiveLocation( BDLocation location) {
    // map view 銷毀後不再處理新接收的位置
        if (location == null || mMapView == null) {
            return;
        }
            MyLocationData locData = new MyLocationData.Builder( )
.accuracy( location.getRadius( ) )
    // 此處設置開發者獲取到的方向信息,順時針 0°~360°
```

```
        .direction(100).latitude(location.getLatitude())
            .longitude(location.getLongitude()).build();
                mBaiduMap.setMyLocationData(locData);
                if(isFirstLoc){
                    isFirstLoc = false;
                    LatLng ll = new LatLng(location.getLatitude(),
location.getLongitude());
                    MapStatus.Builder builder = new MapStatus.Builder();
builder.target(ll).zoom(18.0f);
mBaiduMap.animateMapStatus(
MapStatusUpdateFactory.newMapStatus(builder.build()));
                }
            }
                public void onReceivePoi(BDLocation poiLocation){
                }
        }

        @Override
        protected void onDestroy(){
            super.onDestroy();//在 activity 執行 onDestroy 時執行
mMapView.onDestroy(),實現地圖生命週期管理
            mLocClient.stop();
            mBaiduMap.setMyLocationEnabled(false);    // 關閉定位圖層
            mMapView.onDestroy();
            mMapView = null;
super.onDestroy();
        }
        @Override
        protected void onResume(){
            super.onResume();    //在 activity 執行 onResume 時執行
mMapView.onResume(),實現地圖生命週期管理
            mMapView.onResume();
        }
        @Override
        protected void onPause(){
            super.onPause();    //在 activity 執行 onPause 時執行
mMapView.onPause(),實現地圖生命週期管理
            mMapView.onPause();
        }

    }
```

(3)最後在佈局文件中添加地圖控件即可。

```
<com.baidu.mapapi.map.MapView
    android:id="@+id/bmapView"
android:layout_width="fill_parent"
android:layout_height="fill_parent"
android:clickable="true" />
```

在 Genymotion 模擬器中運行程序,會自動定位到網路所在地址,效果如圖 8.7 所示。

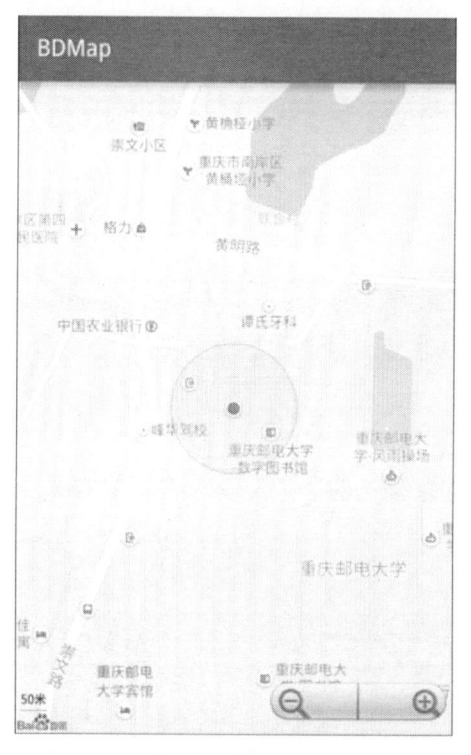

圖 8.7　定位功能

8.4　擴展練習

編寫 Android 項目,實現獲取 GPS 經緯度信息。

8.5　實驗報告

(1)每人一份實驗報告,統一用學校提供的 A4 幅面的實驗報告冊書寫或用 A4

的紙打印。如果打印必須有以下格式的表頭：

實驗課程								
實驗名稱								
實驗時間		學年	學期	週	星期	第	節	
學生姓名		學號			班級			
同組姓名		學號			班級			
實驗地點		設備號			指導教師			

（2）實驗內容的主要結果及對結果的分析。
（3）實驗過程中你所遇到的問題的解決辦法。
（4）心得體會、意見和建議。

8.6　實驗成績考核

（1）考勤占 10%。
（2）相互協作完成實驗任務占 40%。
（3）實驗報告占 50%。

綜合項目一　基於 Android 的計算器

9.1　系統分析

　　計算器是一種在日常生活中應用廣泛的電子產品,無論是在超市商店,還是在辦公室,或是家庭都有著它的身影。基於 Android 的電子計算器的開發更是計算器發展的方向,通過該軟件用戶能夠隨時隨地地進行通用計算,極大地方便使用者。

　　對系統進行可行性分析:

　　1. 要求

　　軟件的功能要符合使用者的基本情況。可以方便地進行加、減、乘、除等運算操作。系統的界面要簡易、快捷,不要有過多的修飾或者不必要的功能。軟件要有一定的安全性和健壯性,不能影響手機系統的運行。

　　2. 目標

　　方便所有對計算器功能有需求的手機用戶。

　　3. 評價尺度

　　軟件要在 5 個工作日內開發完畢並交付使用,用戶需要在 1 個工作日內確認需求文檔,去除其中的不合格因素,程序員需要在用戶確認需求後 3 個工作日內完成系統設計、詳細設計、編碼、測試。最後 1 個工作日在用戶的協助下,完成安裝和反饋修改等工作。

9.2　系統設計

　　1. 系統目標

　　根據用戶對 Android 計算器的使用要求,制定系統目標如下:

　　(1)操作簡單、易於掌握,界面簡潔清爽。

　　(2)方便進行加、減、乘、除等操作。

　　(3)要包含小數點運算和輸入回退功能。

　　(4)能夠進行多次疊加運算。

　　(5)系統運行穩定,不能和手機固有軟件衝突,安全可靠。

　　2. 系統功能結構

　　Android 計算器的功能架構如圖 9.1 所示。

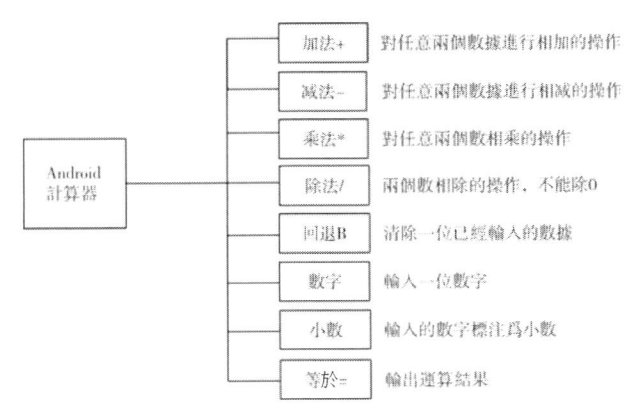

圖 9.1 Android 計算器功能結構圖

3. 系統業務流程

Android 計算器軟件的邏輯流程如圖 9.2 所示。

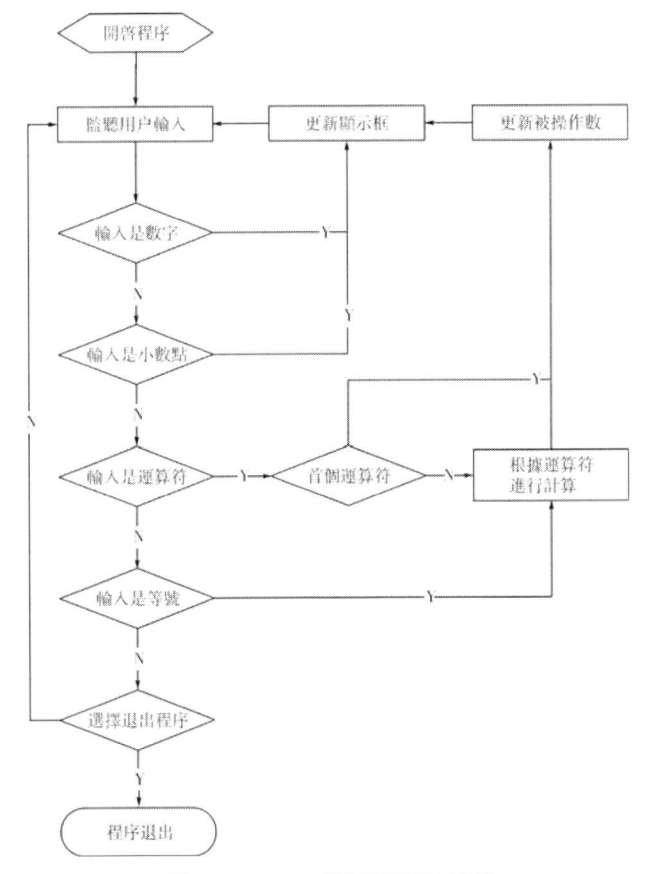

圖 9.2 Android 計算器邏輯流程圖

9.3 系統實施

9.3.1 開發及運行環境

本項目的軟件開發及運行環境如下：
(1)操作系統：Windows7。
(2)開發工具：Android Studio 1.1.0 + Android5.0.1(API 21)。
(3)JDK 環境：Java SE Development Kit(JDK)1.8.0_31。
(4)開發語言：Java,XML。
(5)運行平臺：AVD Nexus S API 21(虛擬機設置)。
(6)分辨率：480*800 hdpi。

9.3.2 項目的創建

Android 計算器系統的項目名稱為 MyCalc，該項目是使用 Android Studio 1.1.0 + Android 5.0.1(API 21)開發的，在 Android Studio 開發環境中創建該項目的步驟如下：

(1)啓動 Android Studio，經過初始化配置之後，會來到如圖 9.3 的界面。在這個頁面我們可以新建項目，也可以導入本地的項目等，左邊可以查看最近打開的項目。這裡直接選擇新建項目。

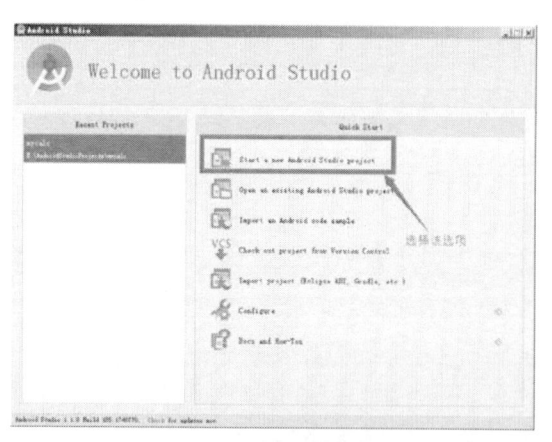

圖 9.3 初始化窗口

(2)單擊新建工程選項後，會彈出 Create New Project 窗體。在該窗體中我們首先輸入項目名稱 myclac，並且設置公司域名為 test.com，然後選擇項目的存儲路徑(如圖9.4所示)，最後點擊「Next」按鈕進入下一步。

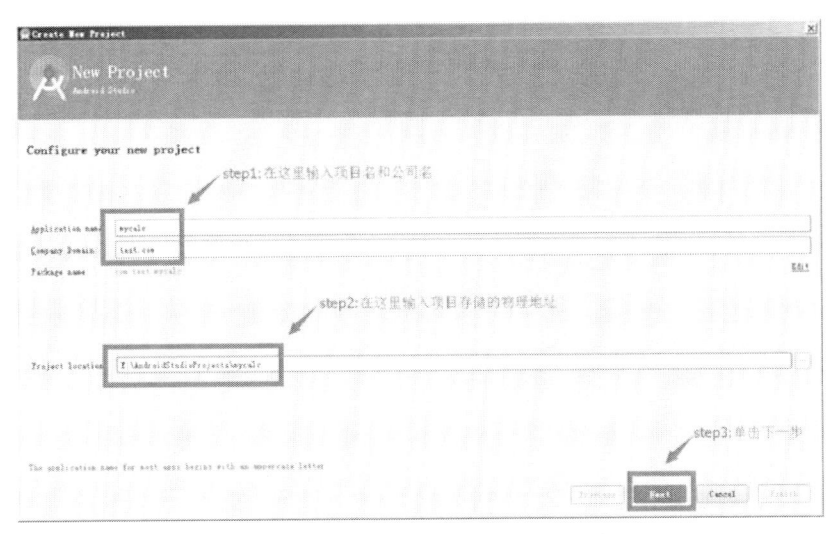

圖 9.4　輸入項目名稱和路徑窗口

（3）Target Android Devices 頁面支持適配 TV，Wear，Glass 等，我們只選擇第一項 Phone 即可，然後選好項目支持的最小 SDK，最後點擊「Next」按鈕進入下一步，如圖 9.5 所示。

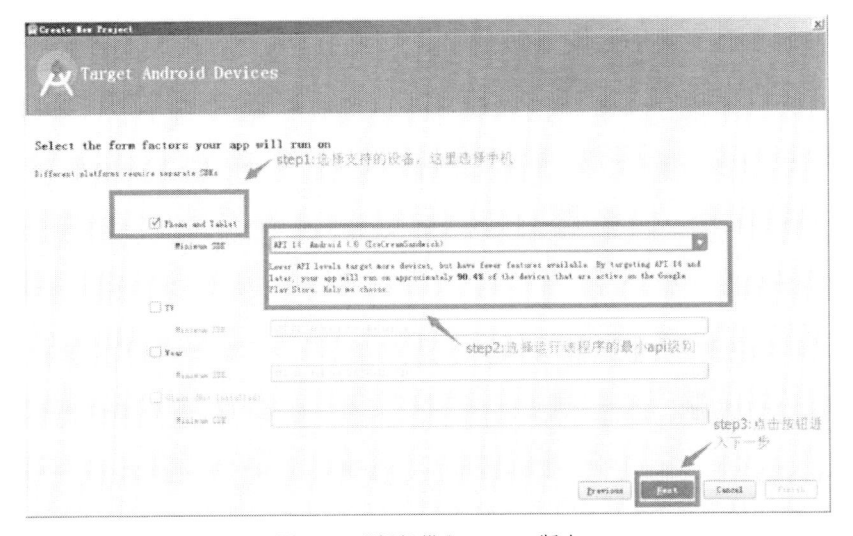

圖 9.5　選擇設備和 Android 版本

（4）在 Add Activity 頁面選擇一個 Activity 模板添加到工程中，和 Eclipse 很像，我們直接選擇 Blank Activity，創建一個「Hello World!」程序，然後點擊「Next」按鈕進入下一步，如圖 9.6 所示。

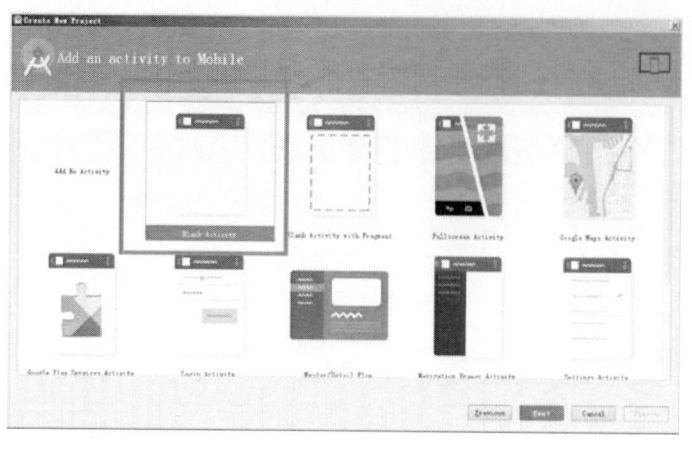

圖 9.6　添加 Activity

（5）在 Customize the Activity 頁面可以設置主 Activity 的名稱、XML 界面配置文件名稱、項目的 Title 和菜單名稱，然後點擊「Finish」按鈕完成工程的創建，如圖 9.7 所示。

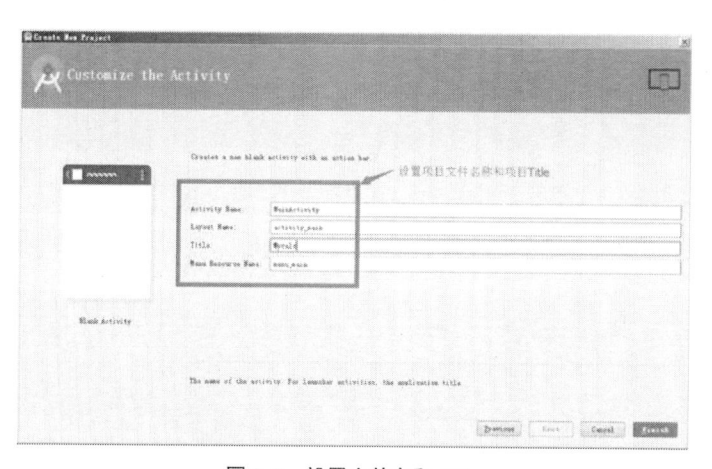

圖 9.7　設置文件名和 Title

9.3.3　計算器主界面實現

主窗體是程序必不可少的元素，是實現人機交互的必有環節。通過主窗體，用戶可以連結到系統的各個子模塊，快速調用系統實現的所有功能。在 Android 計算器中，我們只需求一個主窗體，沒有子窗體的需求，所以所有功能都體現在主窗體上。主窗體以圖標和文字結合的方式顯示計算器的各個功能按鈕，單擊這些按鈕，完成計算器的輸入和計算功能。主窗體運行結果如圖 9.8 所示。

圖 9.8　Android 計算器主窗體

1. 設計窗體佈局 XML 文件

在 Android Studio 的工程中 res\layout\ 目錄下，找到 activity_main.xml 文件，用來作為主窗體佈局的設置文件。在該佈局文件中，添加一個 AbsoluteLayout（絕對佈局管理器）組件，用於顯示功能圖標和圖標上的文本。其代碼如下：

```
<? xml version="1.0" encoding="utf-8" ? >
<AbsoluteLayout android:id="@ +id/widget0"
    android:layout_width="fill_parent"
    android:layout_height="fill_parent"
    xmlns:android="http://schemas.android.com/apk/res/android">
</AbsoluteLayout>
```

2. 添加控件並綁定響應函數

在 AbsoluteLayout（絕對佈局管理器）中，定義一個 TextView 用於顯示中間數字和結果，定義 17 個 Button 用於輸入 0～9 的數字和＋、－、＊、／、＝、.運算符號以及「back」輸入回退按鈕，並用「android:onClick」屬性綁定它們的響應函數。具體代碼如下：

```
<TextView android:id="@ +id/txtResult" android:layout_width="460px"
    android:layout_height="80px" android:background="#ffffffff"
    android:text="0.0" android:textSize="28sp" android:textStyle="bold"
```

            android:textColor="#ff333333" android:layout_x="11dp"
        android:layout_y="16dp" android:gravity="right" android:padding="2px"
/>
    <Button android:id="@+id/btn1" android:layout_width="80px"
        android:layout_height="80px" android:text="1" android:textSize="20sp"
        android:background="#ff5599ff" android:textStyle="bold"
        android:gravity="center" android:layout_x="15dp" android:layout_y="85dp"
        android:onClick="digital_click" />

    <Button android:id="@+id/btn2" android:layout_width="80px"
        android:layout_height="80px" android:text="2" android:textSize="20sp"
        android:background="#ff5599ff" android:textStyle="bold"
        android:gravity="center" android:layout_x="85dp" android:layout_y="85dp"
        android:onClick="digital_click" />

    <Button android:id="@+id/btn3" android:layout_width="80px"
        android:layout_height="80px" android:text="3" android:textSize="20sp"
        android:background="#ff5599ff" android:textStyle="bold"
        android:gravity="center" android:layout_x="155dp" android:layout_y="85dp"
        android:onClick="digital_click" />

    <Button android:id="@+id/btnAdd" android:layout_width="80px"
        android:layout_height="80px" android:text="+" android:textSize="20sp"
        android:textStyle="bold" android:gravity="center" android:layout_x="225dp"
        android:layout_y="85dp" android:onClick="add" />

    <Button android:id="@+id/btn4" android:layout_width="80px"
        android:layout_height="80px" android:text="4" android:textSize="20sp"
        android:background="#ff5599ff" android:textStyle="bold"
        android:gravity="center" android:layout_x="15dp" android:layout_y="155dp"
        android:onClick="digital_click" />

    <Button android:id="@+id/btn5" android:layout_width="80px"

    android:layout_height = "80px" android:text = "5" android:textSize = "20sp"
    android:background = "#ff5599ff" android:textStyle = "bold"
    android:gravity = "center" android:layout_x = "85dp" android:layout_y = "155dp"
    android:onClick = "digital_click" />

<Button android:id = "@+id/btn6" android:layout_width = "80px"
    android:layout_height = "80px" android:text = "6" android:textSize = "20sp"
    android:background = "#ff5599ff" android:textStyle = "bold"
    android:gravity = "center" android:layout_x = "155dp" android:layout_y = "155dp"
    android:onClick = "digital_click" />

<Button android:id = "@+id/btnsub" android:layout_width = "80px"
    android:layout_height = "80px" android:text = "-" android:textSize = "20sp"
    android:textStyle = "bold" android:gravity = "center" android:layout_x = "225dp"
    android:layout_y = "155dp" android:onClick = "sub" />

<Button android:id = "@+id/btn7" android:layout_width = "80px"
    android:layout_height = "80px" android:text = "7" android:textSize = "20sp"
    android:background = "#ff5599ff" android:textStyle = "bold"
    android:gravity = "center" android:layout_x = "15dp" android:layout_y = "225dp"
    android:onClick = "digital_click" />
<Button android:id = "@+id/btn8" android:layout_width = "80px"
    android:layout_height = "80px" android:text = "8" android:textSize = "20sp"
    android:background = "#ff5599ff" android:textStyle = "bold"
    android:gravity = "center" android:layout_x = "85dp" android:layout_y = "225dp"
    android:onClick = "digital_click" />

<Button android:id = "@+id/btn9" android:layout_width = "80px"
    android:layout_height = "80px" android:text = "9" android:textSize = "20sp"
    android:background = "#ff5599ff" android:textStyle = "bold"
    android:gravity = "center" android:layout_x = "155dp" android:layout_y = "225dp"

android:onClick="digital_click" />

<Button android:id="@+id/btnMul" android:layout_width="80px"
　　android:layout_height="80px" android:text=" * " android:textSize="20sp"
　　android:textStyle="bold" android:gravity="center" android:layout_x="225dp"
　　android:layout_y="225dp" android:onClick="mul" />

<Button android:id="@+id/btn0" android:layout_width="130px"
　　android:layout_height="80px" android:text="0" android:textSize="20sp"
　　android:textStyle="bold" android:gravity="center" android:layout_x="15dp"
　　android:layout_y="295dp" android:onClick="digital_click" />

<Button android:id="@+id/btnPoint" android:layout_width="80px"
　　android:layout_height="80px" android:text="." android:textSize="20sp"
　　android:textStyle="bold" android:gravity="center" android:layout_x="155dp"
　　android:layout_y="295dp" android:onClick="point_click" />

<Button android:id="@+id/btnDiv" android:layout_width="80px"
　　android:layout_height="80px" android:text="/" android:textSize="20sp"
　　android:textStyle="bold" android:gravity="center" android:layout_x="225dp"
　　android:layout_y="295dp" android:onClick="div" />
<Button android:id="@+id/btndel" android:layout_width="80px"
　　android:layout_height="85px" android:text="back" android:textSize="15sp"
　　android:textStyle="bold" android:textColor="#ffff0000"
　　android:gravity="center" android:layout_x="15dp" android:layout_y="365dp"
　　android:onClick="del" />

<Button android:id="@+id/btnequ" android:layout_width="290px"
　　android:layout_height="80px" android:text=" = " android:textSize="20sp"
　　android:textStyle="bold" android:gravity="center" android:layout_x="85dp"
　　android:layout_y="365dp" android:onClick="equ" />

9.3.4 計算器邏輯實現

由於計算器本身功能較為單一，實現簡單，所以用一個 Java 文件即可實現對所有功能的編制。建立工程後，對默認的主程序文件，也就是 Android Studio 工程窗體中的 java 文件夾內的「com.test.mycalc」包中的「MainActivity.java」文件進行編輯。該文件所在位置如圖 9.9 所示，下面將該文件分模塊解析。

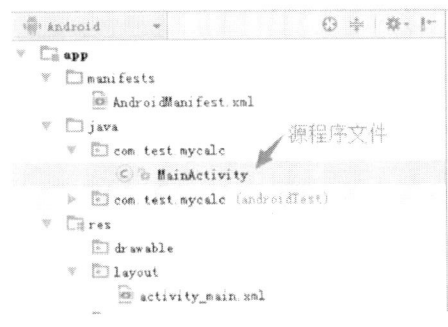

圖 9.9　Android 源程序文件

1. 文件導入的包和類

由於程序需要使用到庫函數中的模塊定義，所以在編寫程序時，需「import」相應的內容來支持程序對該類的使用。例如要使用 Android Studio 中的按鈕元件，就必須導入「Button」類。Android 計算器的文件頭如下：

```
package com.test.mycalc;               //聲明程序所屬包
        import java.text.NumberFormat;        //導入 NumberFormat 類
        import android.app.Activity;          //導入 Activity 類
        import android.os.Bundle;             //導入 Bundle 類
        import android.view.View;             //導入 View 類
        import android.widget.Button;         //導入 Button 類
        import android.widget.TextView;       //導入 TextView 類
```

2. 類的定義

構造函數是程序初次運行時自動調用的函數，一般在這裡對程序中的變量、界面元素等做生成和初始化。Android 計算器的初始化單元如下：

```
public class MainActivity extends Activity {
/** Called when the activity is first created. */

double firstNum = 0;// 第一個輸入的數據
char currentSign = '+';// 記錄第一次輸入的符號
StringBuffer currentNum = new StringBuffer();// 得到 textview 中的數據
boolean isFirstPoint = false;// 第一個數據是否是小數點
```

```
TextView txtResult;// 輸出結果

@Override
public void onCreate(Bundle savedInstanceState) {
    super.onCreate(savedInstanceState);
    setContentView(R.layout.activity_main);

    txtResult = (TextView) findViewById(R.id.txtResult);
}
```

3. 子函數設計

程序需求的各種功能，分別編制子函數實現，可以安排團隊中合理分工。Android 計算器的子函數設計包括以下內容：

（1）輸入框恢復與初始化函數。

用於實現數字數組當前值設置為 0，並清除小數點因素。用於被其他子函數調用。詳細代碼如下：

```
/**
 * 對數據進行初始化
 *
 * @return
 */
public void init() {
    currentNum.delete(0, currentNum.length());// 設置當前 textView 中的值為 0
    isFirstPoint = false;
}
```

（2）數值轉換函數。

用於從輸入框的 string 變量中，獲取輸入的雙精度浮點數。計算函數可以調用。詳細代碼如下：

```
/**
 * 將輸入的數據轉換成 double 類型
 *
 * @return
 */
public double stringToDouble() {
    if (currentNum.length() == 0) // 如果沒有輸入的數據
```

```
        return 0;
    }
    double result = Double.parseDouble(currentNum.toString());
    return result;
}
```

(3)計算函數。

用於實現計算本身,判定進行加、減、乘、除的哪一種運算並執行。運算結果保留兩位小數。詳細代碼如下:

```
/**
 * 進行計算處理
 * @return
 */
public double calcu() {
    double result = 0;
    switch(currentSign) {
        case '+':
            result = firstNum + stringToDouble();
            break;
        case '-':
            result = firstNum - stringToDouble();
            break;
        case '*':
            result = firstNum * stringToDouble();
            break;
        case '/':
            result = firstNum / stringToDouble();
            break;
    }
    //對小數點後的數據進行格式化
    NumberFormat format = NumberFormat.getInstance();
    format.setMaximumFractionDigits(2);
    result = Double.parseDouble(format.format(result));
    return result;
}
```

(4)顯示數據函數。

用於刷新文本框的顯示內容,供其他函數調用。詳細代碼如下:

```
    /**
    * 顯示數據
    */
    public void display( ){

        txtResult.setText( currentNum.toString( ) );
}
```

(5)數字按鍵點擊響應函數。

用於響應 0~9 數字按鈕的點擊動作。自動判斷點擊的按鈕並將該數字添加到文本框，作為輸入的一部分。詳細代碼如下：

```
    /**
    * 處理數據按鈕的點擊
    *
    * @param view
    */
    public void digital_click( View view ) {
        Button btnDigital = ( Button ) view;
        char text = btnDigital.getText( ).charAt( 0 );
        currentNum.append( text );
        display( );
}
```

(6)加法按鍵點擊響應函數。

用於響應加法按鈕的點擊動作。進行加法運算並更新文本框、結果變量、運算符變量並刷新當前輸入數組。詳細代碼如下：

```
    /**
    * 處理加法
    *
    */
    public void add( View view ) {
        double result = calcu( );
        txtResult.setText( String.valueOf( result ) );
        firstNum = result;
        currentSign = '+';
        init( );
}
```

(7)減法按鍵點擊響應函數。

用於響應減法按鈕的點擊動作。進行減法運算並更新文字方塊、結果變數、運算符變數並刷新當前輸入陣列。詳細代碼如下：

```
/**
* 處理減法
*
*/
public void sub(View view) {
    double result = calcu();
    txtResult.setText(String.valueOf(result));
    firstNum = result;
    currentSign = '-';
    init();
}
```

(8)乘法按鍵點擊響應函數。

用於響應乘法按鈕的點擊動作。進行乘法運算並更新文字方塊、結果變數、運算符變數並刷新當前輸入陣列。詳細代碼如下：

```
/**
* 處理乘法
*
*/
public void mul(View view) {
    double result = calcu();
    txtResult.setText(String.valueOf(result));
    firstNum = result;
    currentSign = '*';
    init();
}
```

(9)除法按鍵點擊響應函數。

用於響應除法按鈕的點擊動作。進行除法運算並更新文字方塊、結果變數、運算符變數並刷新當前輸入陣列。詳細代碼如下：

```
/**
* 處理除法
*
*/
```

```java
public void div(View view){
    double result=calcu();
    txtResult.setText(String.valueOf(result));
    firstNum=result;
    currentSign='/';
    init();
}
```

(10)處理等於函數。

用於如果不斷點擊等號按鈕的動作。如果不斷點擊等號按鈕等於累加運算。

```java
/**
 * 處理等於
 *
 */
public void equ(View view){
    double result=calcu();
    txtResult.setText(String.valueOf(result));
    firstNum=result;
    currentSign='+';
    init();
}
```

(11)小數點按鍵點擊響應函數。

用於響應小數點按鈕的點擊動作。不進行運算,判斷是否為第一個小數點,不是則將小數點加入當前輸入數字數組中。詳細代碼如下:

```java
/**
 * 處理小數點
 *
 */
public void point_click(View view){
    if(isFirstPoint){//當地一個數據為小數點時程序返回
        return;
    }
    if(currentNum.length()==0){//當沒有輸入的數據時返回
        return;
    }
    Button btnPoint=(Button)view;
```

```
        char text = btnPoint.getText( ).charAt(0);
        currentNum.append(text);
        isFirstPoint = true;
        display( );
    }
```

(12) 刪除按鍵點擊響應函數。

用於響應刪除按鈕的點擊動作。不進行運算,判斷是否有數字可刪除,有則刪除一位當前輸入數據。詳細代碼如下:

```
/**
 * 處理刪除數據
 */
public void del(View view) {

    if(currentNum.length( ) >= 1)
    {
        currentNum.delete(currentNum.length( ) - 1, currentNum.length( ));
    }
    if(currentNum.length( ) == 0) {

        init( );
        display( );
    }
    txtResult.setText(currentNum);
}
```

9.4 系統運行與測試

界面設計和代碼設計完成後,單擊 Android Studio 開發環境的工具欄中的 ▶ 圖標,或者在菜單欄中的「Run」/「Run」指令,運行該項目,可以調試查看 Android 計算器程序。當我們點擊運行後,會彈出設備選擇窗口,如圖 9.10 所示,這裡可以選擇連接在電腦上的手機設備、運行中的 Android 虛擬機或者新啓動一個 Android 虛擬設備。我們選擇啓動早前在 AVD 創建好的「NexusS」設備。

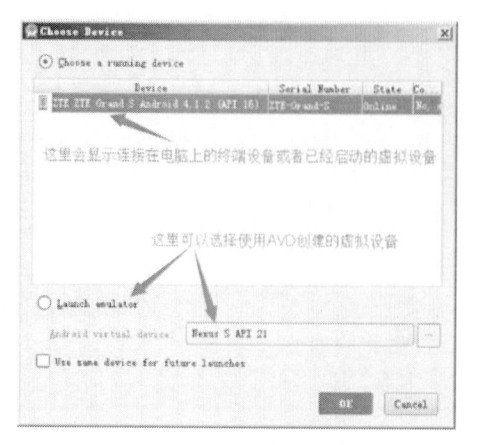

圖 9.10 設備選擇窗體

我們在運行好的窗體中輸入測試數據,「26*3.4=」,結果為「88.4」,如圖 9.11 所示,項目開發完成!

圖 9.11 項目運行結果

綜合項目二　圖形化數字遊戲

10.1　系統分析

手機中的小遊戲是一種在日常生活中應用廣泛的休閒軟件,無論是在超市商店,還是在辦公室,或是家庭都有著它的身影。隨著移動互聯網和智能手機的不斷發展和進步,當今市場上已經出現了數不清的簡單輕鬆的小遊戲,幾乎每一位智能手機的使用者都會在種類繁多的 APP 網站上下載一些小遊戲,數字小遊戲可以很好地讓人們在緊張的工作中放鬆身心。

1. 要求

小遊戲的功能要符合使用者的基本情況,可以方便地進行操作;遊戲的界面要求美觀,有自己的特色;系統的界面要簡易、快捷,不要有過多的修飾或者不必要的功能;遊戲流暢,但不能影響手機系統的運行。

2. 目標

方便所有對休閒娛樂功能有需求的手機用戶。

10.2　系統設計

10.2.1　系統目標

根據用戶對 Android 數字小遊戲的使用要求,制定系統目標如下:
(1)操作簡單、易於掌握,界面簡潔清爽。
(2)方便進行遊戲的觸摸操作。
(3)要包含記分和遊戲終止功能。
(4)系統運行穩定,不能和手機固有軟件衝突,安全可靠。

10.2.2　系統功能結構

Android 數字小遊戲的功能架構如圖 10.1 所示。

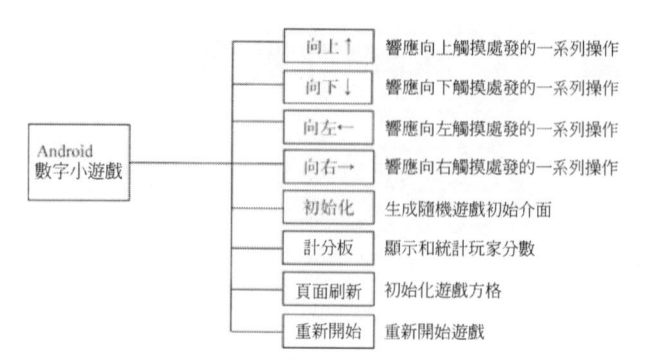

图 10.1　Android 数字小游戏功能结构图

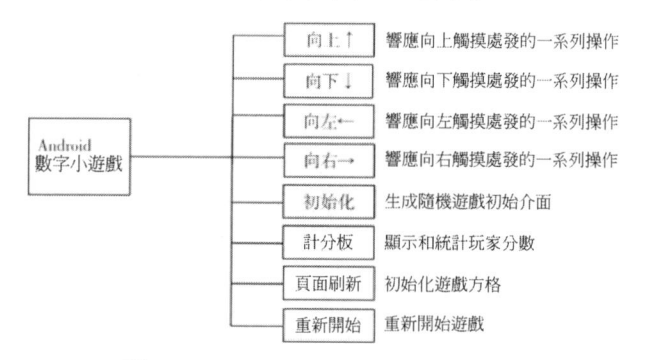

10.2.3　系统业务流程

Android 数字小游戏软件的业务流程如图 10.2 所示。

图 10.2　Android 数字小游戏逻辑流程图

10.3　系统实施

10.3.1　开发及运行环境

本项目的软件开发及运行环境如下：
(1) 操作系统：Windows7。
(2) 开发工具：Android Studio 1.1.0 + Android 4.4.2(API 19)。
(3) JDK 环境：Java SE Development Kit(JDK)1.8.0_31。
(4) 开发语言：Java、XML。
(5) 运行平台：AVD Nexus S API 19(虚拟机设置)。
(6) 分辨率：480 * 800 hdpi。

10.3.2　项目的创建

Android 数字小游戏系统的项目名称为 MiniGame，该项目是使用 Android Studio

1.1.0 + Android 4.4.2（API 19）開發的，在 Android Studio 開發環境中創建該項目的步驟如下：

（1）啓動 Android Studio，經過初始化配置之後，會來到如圖 10.3 所示的界面。在這個頁面我們可以新建項目，也可以導入本地或者 GitHub 上的項目，左邊可以查看最近打開的項目等，這裡直接選擇新建項目。

圖 10.3　初始化窗口

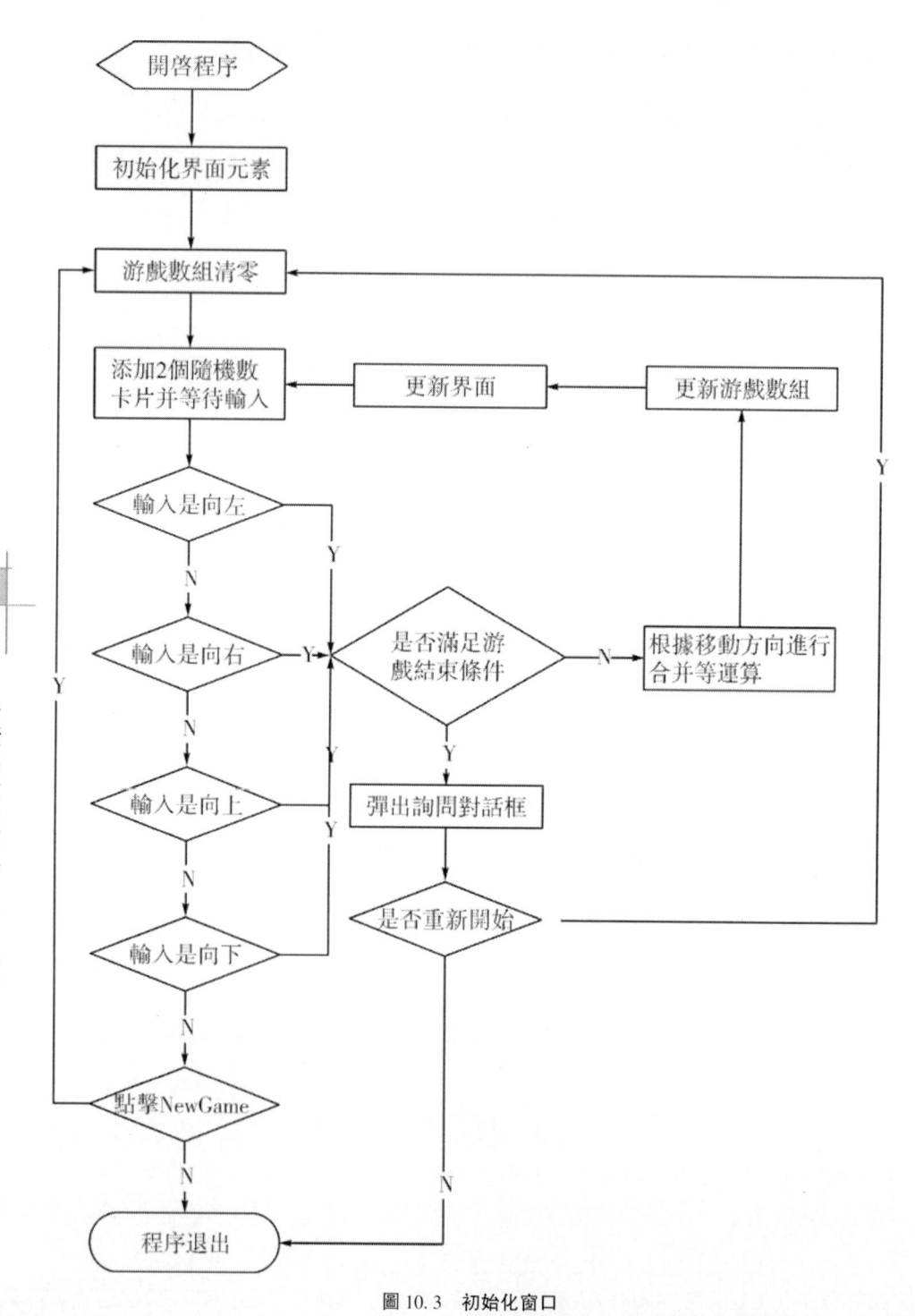

（2）單擊新建工程選項後，會彈出 Create New Project 窗體。在該窗體中我們首先輸入項目名稱 MiniGame，並且設置公司域名為 as.com，然後選擇項目的存儲路徑（如圖 10.4 所示），最後點擊「Next」按鈕進入下一步。

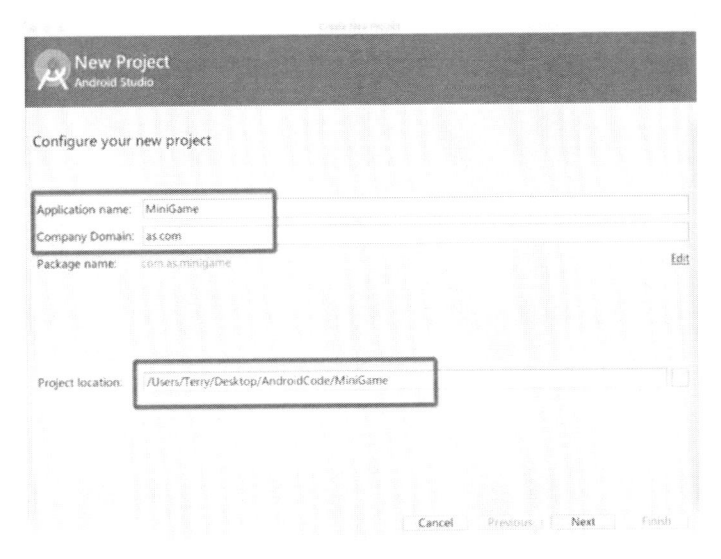

圖 10.4　輸入項目名稱和路徑窗口

（3）Target Android Devices 頁面支持適配 TV、Wear、Glass 等，我們只選擇第一項 Phone 即可，然後選好項目支持的最小 SDK（API 19），最後點擊「Next」按鈕進入下一步，如圖 10.5 所示。

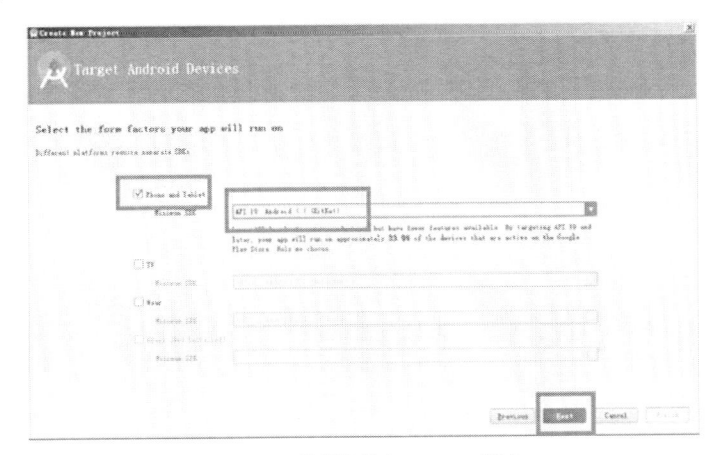

圖 10.5　選擇設備和 Android 版本

（4）在 Add Activity 頁面選擇一個 Activity 模板添加到工程中，和 Eclipse 很像，我們直接選擇 Blank Activity 創建一個「Hello World!」程序，然後點擊「Next」按鈕進入下一步，如圖 10.6 所示。

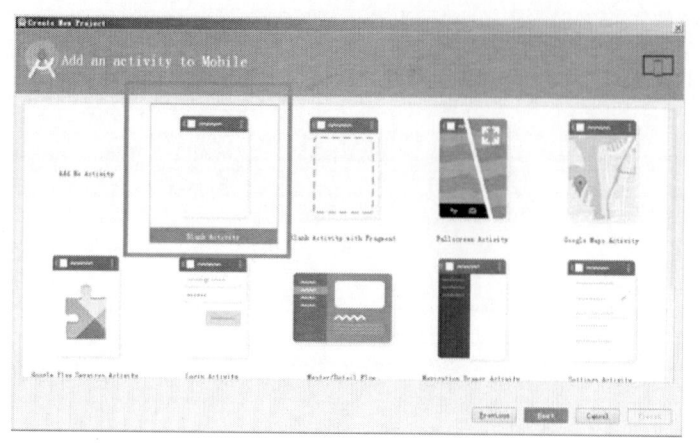

圖 10.6　添加 Activity

（5）在 Customize the Activity 頁面可以設置主 Activity 的名稱、XML 界面配置文件名稱、項目的 Title 和菜單名稱，然後點擊「Finish」按鈕完成工程的創建。

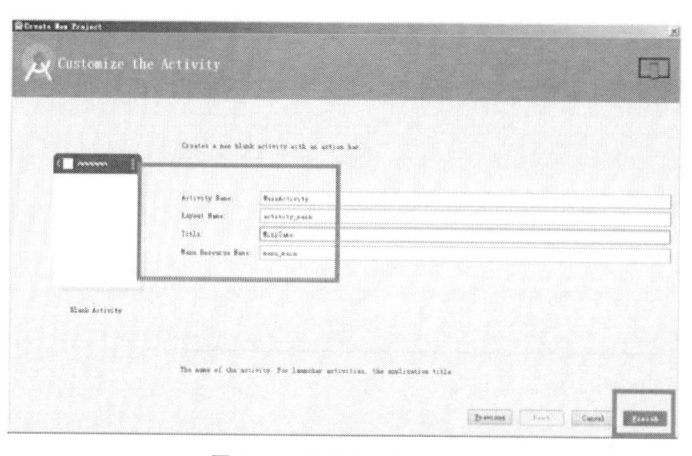

圖 10.7　設置文件名和 Title

10.3.3　項目工程結構

在我們編寫代碼之前，首先要制定好項目的系統文件夾的組織結構目錄，分門別類地管理我們的項目資源，包括不同的窗體、類、數據模型、圖片資源等。這樣不但可以保證系統開發過程的規範性和程序員的可替代性，還有利於保證團隊開發的一致性。創建好系統的文件夾後，在開發的過程中，只需要將我們新創建的類文件或者資源文件、腳本文件等保存到相應的文件夾中即可。Android 數字小遊戲項目所使用的文件夾組織結構如圖 10.8 所示。

圖 10.8　文件夾組織結構

10.3.4　遊戲主界面實現

主窗體是程序必不可少的必要元素,是實現人機交互的必有環節。通過主窗體,用戶可以連結到系統的各個子模塊,快速地調用系統實現的所有功能。在Android數字小遊戲中,我們只需一個主窗體,沒有子窗體的需求,所以所有功能都體現在主窗體上。主窗體以圖標和文字結合的方式顯示各個計數小方塊,通過上下左右四個方向拖動方塊。完成相同數字的相加和記分,從而進行遊戲。主窗體運行結果如圖10.9所示。

圖 10.9　Android 數字小遊戲主窗體

1. 設計窗體佈局 XML 文件

在 Android Studio 的工程中 res\layout\ 目錄下，找到 activity_main.xml 文件，這是用來作為主窗體佈局的設置文件，在該佈局文件中，添加一個 LinearLayout 組件，用於顯示功能圖標和圖標上的文本。其代碼如下：

```xml
<LinearLayout xmlns:android="http://schemas.android.com/apk/res/android"
    xmlns:tools="http://schemas.android.com/tools"
    android:layout_width="match_parent"
    android:layout_height="match_parent"
    android:orientation="vertical"
    tools:context=".MainActivity" >              //設置關聯文件
</LinearLayout>
```

2. 子窗體組件屬性設置

在主窗體中，定義兩個子窗體(佈局)。第一個子窗體定義三個 TextView，用於顯示頂層的提示、動態分數和新開始遊戲按鈕。具體代碼如下：

```xml
<LinearLayout                                    //定義子窗體
    android:layout_width="fill_parent"           //設置控件 ID
    android:layout_height="wrap_content"
    android:orientation="horizontal"
    >
    <TextView                                    //提示文本框
        android:id="@+id/tv_score_text"          //設置控件 ID
        android:layout_width="wrap_content"
        android:layout_height="wrap_content"
        android:textSize="18dp"
        android:textColor="#000000"
        android:padding="8dp"
        android:text="You Score Is:"             //設置初始文本
        />
    <TextView                                    //動態顯示分數文本框
        android:id="@+id/tv_score_show"
        //設置控件屬性
        android:layout_width="wrap_content"
        android:layout_height="wrap_content"
        android:textSize="20dp"
        android:textColor="#000000"
```

```
            android:padding="10dp"
            android:text="0"                              //設置初始文本
            />
        <TextView                           //用做開始「New Game」按鈕的文本框
            android:id="@+id/tv_newgame"                              //設置控件ID
            android:layout_width="wrap_content"
            android:layout_height="wrap_content"
            android:textSize="18dp"
            android:textColor="#000000"
            android:paddingTop="8dp"
            android:paddingLeft="20dp"
            android:text="NEW Game"                         //設置初始文本
            />
    </LinearLayout>
```

第二個子窗體為遊戲子窗體，設置其子窗體大小，具體代碼如下：

```
    <com.as.minigame.GameView                              //關聯到GameView 類
        android:id="@+id/gv_show"                   //設置控件ID
        android:layout_width="fill_parent"
        android:layout_height="fill_parent"
        android:layout_weight="1"
        >
    </com.as.minigame.GameView>
```

10.3.5 數字小遊戲邏輯實現

小遊戲涉及圖像的更新和分數統計，又要兼顧總體運行。為便於調試和修改，也為了更好地分工，所以採用三個 Java 類文件實現對所有功能的編制。建立工程後，除默認的主程序文件，也就是 Android Studio 工程窗體中的 Java 文件夾內的「com.as.minigame」包中的「MainActivity.java」文件外。另外再創建兩個「Java Class」，分別為「Card.java」和「GameView.java」。文件所在位置如圖 10.10 所示，下面對三個文件分模塊解析。

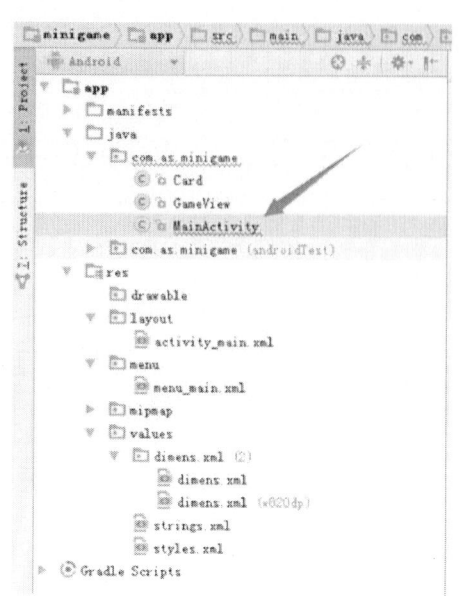

圖 10.10 Android 源程序文件

1. 實現主程序 MainActivity

Android 數字小遊戲工程中的「MainActivity.java」文件主要包含了程序主窗體類的定義，以及各個元件的定義和指定窗體元素與響應函數的關聯。下面詳細解析該類的各個模塊：

（1）文件導入的包和類。

由於該類需要使用到庫函數中的模塊定義，所以在編寫程序時，需「import」相應的內容來支持程序對該類的使用。例如要使用 Android Studio 中的「文本」元件，就必須導入「TextView」類。Android 數字小遊戲的文件頭如下：

```
package com.as.minigame;                                //設置包名

import java.util.Timer;
import java.util.TimerTask;
import android.os.Bundle;
import android.os.Handler;
import android.os.Message;
import android.app.Activity;
import android.view.View;
import android.view.View.OnClickListener;
import android.widget.TextView;
```

(2)類的定義。

Android 數字小遊戲的初始化單元如下:

```
public class MainActivity extends Activity {
    TextView score_show;
    GameView gv;
    TextView new_game;
    Handler handler = new Handler(){

        @Override
        public void handleMessage(Message msg){
            // TODO Auto-generated method stub
            super.handleMessage(msg);
            int num = msg.arg1;
            score_show.setText(num+" ");




        }

    };

    protected void onCreate(Bundle savedInstanceState){
        super.onCreate(savedInstanceState);
        setContentView(R.layout.activity_main);
        score_show = (TextView) findViewById(R.id.tv_score_show);
        gv = (GameView) findViewById(R.id.gv_show);
        new_game = (TextView) findViewById(R.id.tv_newgame);
        new_game.setOnClickListener(new OnClickListener(){

            @Override
            public void onClick(View arg0){
                gv.GameStart();
                gv.score = 0;


            }
        });
```

(3)程序初始化工作。

程序需求的各種初始化,可在主類中編制子函數實現,Android 數字小遊戲的子函數設計包括以下內容:

設置計時器 timer 並使用 timer 定時傳遞包含分數信息的 message,以便界面定時刷新分數。詳細代碼如下：

```
    Timer timer = new Timer();
timer.schedule(new TimerTask() {

    @Override
    public void run() {
        Message msg = new Message();
        msg.arg1 = gv.score;
        handler.sendMessage(msg);
    }
}, 80, 150);
score_show.setText(100+"");
```

2. 實現卡片文件 Card.java

Android 數字小遊戲工程中的「Card.java」文件主要包含了遊戲界面的定義，劃分出遊戲需要的方格，以及各個方塊的位置、大小、顏色、初始文字等。下面詳細解析該類的各個模塊：

(1) 文件導入的包和類。

由於該類需要使用庫函數中的模塊定義，所以在編寫程序時，需「import」相應的內容來支持程序對該類的使用。Card.java 文件頭如下：

```
package com.as.minigame;                    //文件所屬包

import android.content.Context;
import android.view.Gravity;
import android.widget.FrameLayout;
import android.widget.TextView;
```

(2) 實現 Card.java 文件。

Card.java 的初始化單元如下：

```
public class Card extends FrameLayout {

    private TextView text;
    private int number = 0;
    public int getNumber() {
        return number;
    }
```

```java
public void setNumber(int number){
    this.number = number;
    if(number<2){
        text.setText("");
    }else{


        if(number>=64){
            text.setTextColor(0xffffff00);
        }else{
            text.setTextColor(0xff000000);
        }
        text.setText(number+"");
    }


}


public Card(Context context){
    super(context);
    // TODO Auto-generated constructor stub
    text = new TextView(context);
    text.setTextSize(28);
    text.setBackgroundColor(0x9966cccc);
    text.setGravity(Gravity.CENTER);
    LayoutParams params = new LayoutParams(-1,-1);
    params.setMargins(10, 10, 0, 0);
    addView(text, params);


}
}
```

3. 實現遊戲交互文件 GameView.java

Android 數字小遊戲工程中的「GameView.java」文件主要包含了遊戲交互的響應、分數計算、遊戲結束的判定等遊戲的主要功能的實現，以及遊戲界面的初始化和遊戲每一步結果的實現和反饋。下面詳細解析該類的各個模塊：

(1) 文件導入的包和類。

由於該類需要使用到庫函數中的模塊定義，所以在編寫程序時，需「import」相應的內容來支持程序中對該類的使用。GameView.java 文件頭如下：

```java
package com.as.minigame;              //文件所屬包

import java.util.ArrayList;           //導入該類定義所需的其他類和函數
import java.util.List;
import java.util.Random;
import android.app.AlertDialog;
import android.content.Context;
import android.content.DialogInterface;
import android.graphics.Point;
import android.util.AttributeSet;
import android.view.MotionEvent;
import android.view.View;
import android.widget.GridLayout;
```

(2) 實現 GameView.java。

GameView.java 的初始化單元如下：

```java
public class GameView extends GridLayout {

    private Card cards[][] = new Card[4][4];
    private List<Point> emptyCards = new ArrayList<Point>();
    Random rd = new Random();
    int score = 0;

    public GameView(Context context) {
        super(context);
        // TODO Auto-generated constructor stub
        initGame();
    }

    public GameView(Context context, AttributeSet attrs) {
        super(context, attrs);
        // TODO Auto-generated constructor stub
        initGame();
    }
}
```

```java
public GameView(Context context, AttributeSet attrs, int defStyle) {
    super(context, attrs, defStyle);
    // TODO Auto-generated constructor stub
    initGame();
}
}
```

(3)遊戲初始化函數。

該函數的主要功能為設定遊戲的輸入監聽,並定義如何判斷用戶的遊戲輸入,關聯到相應的處理函數。具體代碼如下:

```java
private void initGame() {
    setColumnCount(4);
    setBackgroundColor(0xffffcccc);

    setOnTouchListener(new OnTouchListener() {

        private float startX, startY;
        private float offsetX, offsetY;

        @Override
        public boolean onTouch(View v, MotionEvent event) {
            // TODO Auto-generated method stub

            switch (event.getAction()) {
                case MotionEvent.ACTION_DOWN:
                    startX = event.getX();
                    startY = event.getY();
                    break;
                case MotionEvent.ACTION_UP:

                    Gameover();

                    offsetX = event.getX() - startX;
                    offsetY = event.getY() - startY;
                    if (Math.abs(offsetX) > Math.abs(offsetY)) {
                        if (offsetX < -3) {
                            moveLeft();
```

```
                    System.out.println("----左");
                } else if ( offsetX > 3 ) {
                    moveRight( );
                    System.out.println("----右");
                }
            } else {
                if ( offsetY < -3 ) {
                    moveUp( );
                    System.out.println("----上");
                } else if ( offsetY > 3 ) {
                    moveDown( );
                    System.out.println("----下");
                }

            }

            break;
        default:
            break;
        }
        return true;
    }
});
```

(4) 遊戲觸摸響應函數。

該工程一共定義了上下左右四個觸摸響應函數,該函數的主要功能為響應用戶的輸入,將結果數組按照輸入的方向進行更新並創建新的隨機數卡牌。具體代碼如下:

```
private void moveRight( ) {

    boolean flage = false;
    for ( int y = 0; y < 4; y++) {
        for ( int x = 3; x >= 0; x--) {
            for ( int x1 = x - 1; x1 >= 0; x1--) {
```

```java
                    // 當同一行為空,不需處理
                    if (cards[x1][y].getNumber() > 0) {
                        if (cards[x][y].getNumber() < 2) {
                            // 將前一張卡片的值移動到當前卡片
                            cards[x][y].setNumber(cards[x1][y].getNumber());
                            cards[x1][y].setNumber(0);
                            x++;
                            flage = true;
                            score += 2;
                        } else if (cards[x][y].getNumber() == cards[x1][y].getNumber()) {
                            cards[x][y].setNumber(cards[x][y].getNumber() * 2);
                            score += cards[x][y].getNumber();
                            cards[x1][y].setNumber(0);
                            flage = true;
                        }
                        break;
                    }
                }
            }
        }
    }
    if (flage) {
        creatRandomCard();
    }
}

private void moveLeft() {

    boolean flage = false;
    for (int y = 0; y < 4; y++) {
        for (int x = 0; x < 4; x++) {
            for (int x1 = x + 1; x1 < 4; x1++) {
```

```
                    // 當同一行為空,不需處理
                    if (cards[x1][y].getNumber() > 0) {
                        if (cards[x][y].getNumber() < 2) {
                            // 將前一張卡片的值移動到當前卡片
                            cards[x][y].setNumber(cards[x1][y].getNumber
()); cards[x1][y].setNumber(0);
                            x--;
                            flage = true;
                            score += 2;
                        } else if (cards[x][y].getNumber() == cards[x1][y]
                                .getNumber()) {
                            cards[x][y].setNumber(cards[x][y].getNumber()
* 2);
                            score += cards[x][y].getNumber();
                            cards[x1][y].setNumber(0);
                            flage = true;
                        }
                        break;
                    }
                }
            }
        }

        if (flage) {
            creatRandomCard();
        }
}

private void moveDown() {
    boolean flage = false;
    for (int x = 0; x < 4; x++) {
        for (int y = 3; y >= 0; y--) {
            for (int y1 = y - 1; y1 >= 0; y1--) {
                // 當同一行為空,不需處理
                if (cards[x][y1].getNumber() > 0) {
                    if (cards[x][y].getNumber() < 2) {
```

```
                    // 將前一張卡片的值移動到當前卡片
                    cards[x][y].setNumber(cards[x][y1].getNumber());
                    cards[x][y1].setNumber(0);
                    y++;
                    flage = true;
                    score += 2;
                } else if (cards[x][y].getNumber() == cards[x][y1].getNumber()) {
                    cards[x][y].setNumber(cards[x][y].getNumber() * 2);
                    score += cards[x][y].getNumber();
                    cards[x][y1].setNumber(0);
                    flage = true;
                }
                break;
            }
        }
    }
}
if (flage) {
    creatRandomCard();
}
}

private void moveUp() {
    boolean flage = false;
    for (int x = 0; x < 4; x++) {
        for (int y = 0; y < 4; y++) {
            for (int y1 = y + 1; y1 < 4; y1++) {
                if (cards[x][y1].getNumber() > 0) {
                    if (cards[x][y].getNumber() < 2) {
```

```
                        cards[x][y].setNumber(cards[x][y1].getNumber
());
                        cards[x][y1].setNumber(0);
                        y--;
                        flage = true;
                        score+=2;
                    } else if (cards[x][y].getNumber() = = cards[x][y1]
                        .getNumber()) {
                        cards[x][y].setNumber(cards[x][y].getNumber()
* 2);
                        score +=cards[x][y].getNumber();
                        cards[x][y1].setNumber(0);
                        flage = true;
                    }
                    break;
                }
            }
        }
    }
    if (flage) {
        creatRandomCard();
    }
}
```

(5)遊戲結束判定及處理函數。

根據遊戲設定的規則,在所有卡片皆不為零,而且臨近卡牌分數都不相同、不能合併的情況下,判斷為遊戲結束。遊戲結束時,彈出對話框詢問是否開始新遊戲。具體代碼如下:

```
private void Gameover() {
    boolean OverGame=true;
    for (int y = 0; y < 4; y++) {
        for (int x = 0; x < 4; x++) {
            if(cards[x][y].getNumber() <= 0 ||
                (x >0 && cards[x][y].getNumber() = = cards[x-1]
[y].getNumber())||
```

```
                    (x<3 && cards[x][y].getNumber()==cards[x+1]
[y].getNumber())||
                    (y>0 && cards[x][y].getNumber()==cards[x][y-1].
getNumber())||
                    (y<3 && cards[x][y].getNumber()==cards[x][y+1].
getNumber())
                    ){
                    OverGame=false;
                }
            }
        }
        if(OverGame){
            new AlertDialog.Builder(getContext()).setTitle("hi").setMessage("a-gain").
                    setPositiveButton("yes",new AlertDialog.OnClickListener(){

                        @Override
                        public void onClick(DialogInterface dialog, int which){
                            // TODO Auto-generated method stub

                            GameStart();
                            score = 0;
                        }
                    }).setNegativeButton("No", null).show();
        }
    }
}
```

（6）其他函数。

为满足程序需要，还需要定义其他函数来满足程序的功能。其中包括卡片的添加函数「AddCard(int width, int height)」，界面尺寸变化的响应函数（当创建时也会调用）「onSizeChanged(int w, int h, int oldw, int oldh)」，创建随机有分卡片的「creatRandomCard()」函数，游戏数组初始化的开始函数「GameStart()」。具体代码如下：

```
private void AddCard(int width, int height){

    Card c;
    for (int y = 0; y < 4; y++){

        for (int x = 0; x < 4; x++){
```

```
            c = new Card(getContext());
            cards[x][y] = c;
            c.setNumber(0);
            addView(c, width, height);
        }
    }
}

@Override
protected void onSizeChanged(int w, int h, int oldw, int oldh) {
    // TODO Auto-generated method stub
    super.onSizeChanged(w, h, oldw, oldh);
    int width = (w - 10) / 4;
    AddCard(width, width);
    GameStart();
}

private void creatRandomCard() {

    emptyCards.clear();
    for (int y = 0; y < 4; y++) {
        for (int x = 0; x < 4; x++) {
            if (cards[x][y].getNumber() < 2) {
                Point point = new Point(x, y);
                emptyCards.add(point);
            }
        }

    }
    int selat = rd.nextInt(emptyCards.size());
    Point p = emptyCards.get(selat);
    emptyCards.remove(selat);
    int number = 0;
    if (rd.nextInt(10) > 4) {
        number = 4;
    } else
        number = 2;
```

```
        cards[p.x][p.y].setNumber(number);
    }

    public void GameStart() {
        for (int y = 0; y < 4; y++) {
            for (int x = 0; x < 4; x++) {
                cards[x][y].setNumber(0);
            }
        }
        creatRandomCard();
        creatRandomCard();
    }
```

10.4 系統運行與測試

界面設計和代碼設計完成後,單擊Android Studio開發環境的工具欄中的「▶」圖標,或者在菜單欄中的「Run」/「Run」指令,運行該項目,可以調試查看Android數字小遊戲程序。當我們點擊運行後,會彈出設備選擇窗口,如圖10.11所示,這裡可以選擇連接在電腦上的手機設備、運行中的Android虛擬機或者新啓動一個Android虛擬設備。我們選擇啓動早前在AVD創建好的「NexusS」設備。

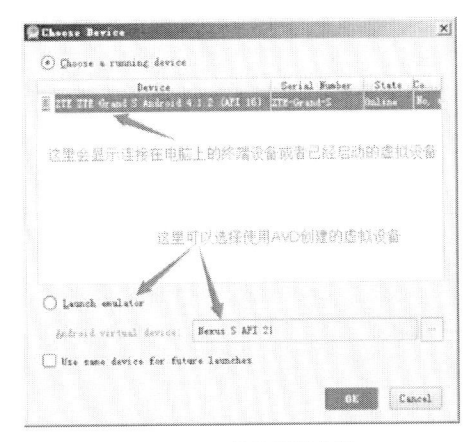

圖10.11 設備選擇窗體

遊戲運行效果如圖10.12所示。

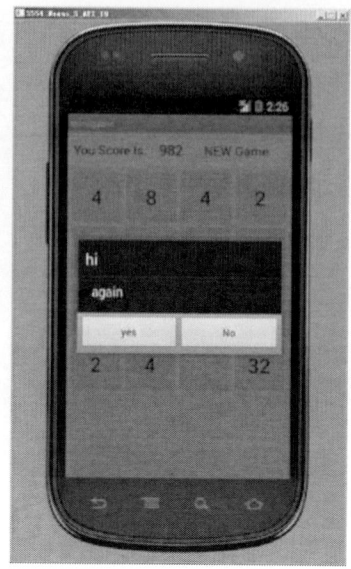

圖 10.12　項目運行結果——遊戲開始及結束

國家圖書館出版品預行編目(CIP)資料

手機終端軟件開發實驗Android版 / 羅文龍 主編. -- 第一版.
-- 臺北市：崧燁文化，2018.08

面 ； 公分

ISBN 978-957-681-426-6(平裝)

1.系統程式 2.電腦程式設計

312.52　　　　107012247

書　名：手機終端軟件開發實驗Android版
作　者：羅文龍 主編
發行人：黃振庭
出版者：崧燁文化事業有限公司
發行者：崧燁文化事業有限公司
E-mail：sonbookservice@gmail.com
粉絲頁　　　　　　網　址：
地　址：台北市中正區重慶南路一段六十一號八樓815室
8F.-815, No.61, Sec. 1, Chongqing S. Rd., Zhongzheng
Dist., Taipei City 100, Taiwan (R.O.C.)
電　話：(02)2370-3310　傳　真：(02) 2370-3210
總經銷：紅螞蟻圖書有限公司
地　址：台北市內湖區舊宗路二段121巷19號
電　話：02-2795-3656　傳真:02-2795-4100　網址：
印　刷：京峯彩色印刷有限公司（京峰數位）

　　本書版權為西南財經大學出版社所有授權崧博出版事業股份有限公司獨家發行電子書繁體字版。若有其他相關權利需授權請與西南財經大學出版社聯繫，經本公司授權後方得行使相關權利。

定價：350 元
發行日期：2018 年 8 月第一版
◎ 本書以POD印製發行